Mathematical modelling techniques

R Aris

University of Minnesota

Mathematical modelling techniques

Pitman

LONDON · SAN FRANCISCO · MELBOURNE

PITMAN PUBLISHING LIMITED
39 Parker Street, London WC2B 5PB

FEARON-PITMAN PUBLISHERS INC.
6 Davis Drive, Belmont, California 94002, USA

Associated Companies
Copp Clark Ltd, Toronto
Pitman Publishing New Zealand Ltd, Wellington
Pitman Publishing Pty Ltd, Melbourne

First Published 1978

British Library Cataloguing in Publication Data

Aris, Rutherford
    Mathematical modelling. - (Research notes
    in mathematics; no.24).
    1. Mathematical models
    I. Title   II. Series
    511'.8   QA402

    ISBN 0-273-08413-5

Reproduced and printed by photolithography
in Great Britain at Biddles of Guildford

ISBN 0 273 08413 5

To my friends at the California Institute of Technology, than whom there are no hosts more gracious nor colleagues more stimulating. Floreat Schola Artium Utiliorum Californiana.

Preface

"Par ma foi! il y a plus de quarante ans que
je dis de la prose sans que j'en susse rien,
et je vous suis le plus obligé du monde de
m'avoir appris cela."

M. Jourdain in Moliere's "Le Bourgeois
Gentilhomme" (Act. II Sc. IV)

The original title under which these notes were written--"Notes toward the
definition of the craft of mathematical modelling"--was somewhat long-winded
and perhaps, by reason of its allusion, a shade pretentious. It had however
the merit of greater precision and conveyed the tentative spirit in which
these notes are put forward for the criticism of a larger public. The whole
activity of mathematical modelling has blossomed forth into such a multitude
of areas in the last few years (witness a 1st International Conference with 2646
pages of Proceedings [199]) that there is indeed a need to define it in the
sense of seeking out its boundary and exploring its interior as well as of
discovering its structure and essential nature. The time is not yet ripe for
a magisterial survey, which would in any case demand an abler pen than mine,
but I believe it can be approached from the angle of craftsmanship. It is a
commonplace in educational circles that it is comparatively easy to teach the
method of solution of a standard mathematical equations, but much harder to
communicate the ability to formulate the equations adequately and economi-
cally. With the notable exception of Lin and Segel's [223], and Haberman's book
[213] and the papers of Hammersley [80,81,82] few publications pay much

attention to the little things that the experienced mathematical modeller does, almost by instinct. It would therefore seem to be worthwhile to try and set down some of these notions in the interests of the craft and with the hope that it will stimulate further discussion and development. There is manifestly a danger here, for it may be only the MM. Jourdains who will be vastly excited to learn that they have been talking prose all their lives. Nevertheless I hope that some of my peers and betters will find the subject worthy of their attention.

Later iterations of this effort will demand a wealth of examples drawn from all branches of the physical and social sciences. In this first attempt I have chosen three physical examples to serve the illustration of many points. These examples—the packed bed, the chromatographic column and the stirred tank—are given in detail in the appendices. They are in some sense fold-out maps to the text though they cannot be presented as such. (Each has its own nomenclature which is listed at the end of its discussion; the nomenclature for other examples is introduced in situ.) These examples and those introduced at various points of the text are often connected with the mathematical theory of chemical reactors. I make no apology for this; the field is a rich one that has stimulated some of the work of the best applied mathematicians who have used a reactor like a stalking-horse under cover of which to shoot their wit. Its problems are challenging, yet from the modelling point of view they do not demand any great knowledge of chemistry or of engineering and so are accessible to all.

Many of the notions I have advanced here and the order I have tried to impose on the subject are quite tentative and I shall appreciate any comments and criticism. I have already benefitted from interaction with colleagues, both faculty and students, at Caltech and it is one of the virtues of

Pitman's Research Notes for Mathematics series that it quickly submits ideas to a wider public.

To the California Institute of Technology I am vastly indebted for a term as a Sherman Fairchild Distinguished Scholar in the fall of 1976, under conditions of such generous hospitality that the fruits of such a tenure can never be worthy of the opportunity. At the risk of overlooking someone, I would like to thank in particular (and in alphabetical order) Cohen, Gavalas, Keller, Pings, Seinfeld and Weinberg. Yolande Johnson did a splendid job of the first draft of these notes that was prepared at Caltech, being helped by Sharon ViGario in the last minute rush. The final version was typed by Shirley Tabis who met the exacting requirements of camera readiness with great skill and dispatch. I am most grateful to all of them.

Contents

What is a model?

ὦ Μένανδρε καὶ βίε
πότερος ἄρ' ὑμῶν πότερον ἀπεμιμήσατο

*attr. to Aristophanes of Byzantium
Syrianus in Hermog. (Rabe ij 23)*

(O Menander, o life, which of you has imitated the other?)

1.1 The idea of a mathematical model and its relationship to other uses of
of the word.

In these notes the term 'mathematical model'--usually abbreviated to
'model'--will be used for any complete and consistent set of mathematical
equations which is thought to correspond to some other entity, its prototype.
The prototype may be a physical, biological, social, psychological or conceptual entity, perhaps even another mathematical model, though in detailed
examples we shall be concerned with a few physico-chemical systems.

Being derived from 'modus' (a measure) the word 'model' implies a change
of scale in its representation and only later in its history did it acquire
the meaning of a type of design, as in Cromwell's New Model Army (1645).
Still later (1788) came the complacent overtones of the exemplar that Gilbert
was to use so effectively for his modern major general, while it is the first
years of this century before fashion became so self-conscious as to claim
its own models and make possible Kaplan's double entendre (see quotation at
head of Ch. 5). In the sense that we are seeking a different scale of

1

thought or mode of understanding we are using the word in its older meaning. However, the word model (without the adjective 'mathematical') has been and is used in a number of senses both by philosophers and scientists as merely glancing through the titles of the bibliography will suggest. Thus Apostel [7] distinguishes nine motivations underlying the use of models ranging from the replacement of a theory-less domain of facts by another for which a theory is known (e.g., network theory as a model for neurological phenomena) to the use of a model as a bridge between theory and observation. Suppes in the same volume [169] maintains that the logician's concept of a model is the same in the empirical sciences as in mathematics though the use to which they are put is different. The logician's definition he takes from Tarski [172] as: "a possible realization in which all valid sentences of a theory T are satisfied is called a model of T". This is a non-linguistic entity in which a theory is satisfied and Suppes draws attention to the confusion that can arise when model is used for the set of assumptions underlying a theory, i.e. the linguistic structure which is axiomatized. In our context this suggests that we might usefully distinguish between the prototype (i.e. the physical entity or system being modelled), the precursive assumptions or what the logicians call the theory of the model (i.e. the precise statement of the assumptions of axioms) and the model itself (i.e. the scheme of equations).

The idea of a change of scale which inheres in the notion of a model through its etymology can be variously interpreted. In so far as the prototype is a physical or natural object, the mathematical model represents a change on the scale of abstraction. Certain particularities will have been removed and simplifications made in obtaining the model. For this reason some hard-headed, practical-minded folk seem to regard the model as less "real" than the prototype. However from the logical point of view the prototype is in fact a realization in which the valid sentences of the mathematical model are to some degree satisfied. One could say that the prototype is a model of equations and the two enjoy the happy reciprocality of Menander and life.

The purpose for which a model is constructed should not be taken for granted but, at any rate initially, needs to be made explicit. Apostel (loc. cit.) recognizes this in his formalization of the modelling relationship $R(S,P,M,T)$, which he describes as the subject S taking, in view of a purpose P, the entity M as a model for the prototype T. J. Maynard Smith [165] uses the notion of purpose to distinguish mathematical descriptions of ecological systems made for practical purposes from those whose purpose is theoretical. The former he calls 'simulations' and points out that their value increases with the amount of particular detail that they incorporate. Thus in trying to predict the population of a pest the peculiarities of its propagation and predilections of its predators would be incorporated in the model with all the specific detail that could be mustered. But ecological theory also seeks to make general statements about the population growth that will discern the broad influence of the several factors that come into play. The mathematical descriptions that serve such theoretical purposes should include as little detail as possible but preserve the broad outline of the problem. These descriptions are called 'models' by Smith, who also comments on a remark of Levins [114] that the valuable results from such models are the indications, not of what is common to all species or systems, but of the differences between species of systems.

Hesse [92] in her excellent little monograph "Models and Analogies in Science" distinguished two basic meanings of the word 'model' as it is used in physics and Leatherdale in a very comprehensive discussion of "The Role of Analogy, Model and Metaphor in Science" has at least four. They stem from the methods of "physical analogy" introduced by Kelvin and Maxwell who used the partial resemblance between the laws of two sciences to make one serve as illustrator of the other. In the hands of 19th century English physicists these often took the form of the mechanical analogues that evoked Duhem's famous passage of Gallic ire and irony. Duhem [56] had in mind that a physical theory should be a purely deductive structure from a small number of rather general hypotheses, but Campbell [41] claimed that this logical consistency was not enough and that links to or analogies with already established laws must be maintained. Leatherdale's four types are the formal and informal variants of Hesse's two. Her "model$_1$" is a copy, albeit imperfect, with certain features that are positively analogous and certain which are neutral but shorn of all features which are known to be negatively analogous, i.e. definitely dissimilar to the prototype. Her "model$_2$" is the copy with all its features, good, bad and indifferent. Thus billiard balls in motion, colored and shiny, are a model$_2$ for kinetic theory, whilst billiard balls in motion obeying perfectly the laws of mechanics but bereft of their colour, shine and all other non-molecular properties constitute a model$_1$. It is the natural analogies (i.e. the features as yet of unknown relevance) that are regarded by Campbell as the growing points of a theory. In these terms a mathematical model would presumably be a formal model$_1$.

Brodbeck [35], in the context of the social sciences, stresses the aspect of isomorphism and reciprocality when she defines a model by saying that if the laws of one theory have the same form as the laws of another theory, then one may be said to be a model for the other. There remains, of course, the problem of determining whether the two sets of laws are isomorphic.

4

Brodbeck further distinguishes between two empirical theories as models one

of the other and the situation when one theory is an "arithmetical

structure". She then goes on to describe three meanings of the term

mathematical model according as the modelling theory is (a) any quantified

empirical theory, (b) an arithmetic structure or (c) a mere formalization in

which descriptive terms are given symbols in the attempt to lay bare the

axioms or otherwise to examine the structure of the theory. If arithmetical

is interpreted with suitable breadth we are clearly concerned in these notes

with sense (b).

It is obviously inappropriate in the present context to try to survey all

the senses in which the word has been used, among which there is no lack of

confusion. A more formal version of the definition of a (mathematical) model

that we started with might be as follows: a system of equations, Σ, is said

to be a model of the prototypical system, S, if it is formulated to express

the laws of S and its solution is intended to represent some aspect of the

behavior of S. This is vague enough in all conscience, but the isomorphism

is never exact and we deny the name of modelling to the less successful

efforts of the game. Rather, we should try and find out what constitutes a

good or bad model.

It scarcely needs to be added that we shall not raise the old red herring

about the model being less "real" than the prototype. Tolkien [178] has

reminded us of the failure of the expression "real life" to live up to

academic standards. "The notion", he remarks, "that motor cars are more

'alive' than, say, centaurs or dragons is curious; that they are more 'real'

than, say, horses is pathetically absurd".

The mention of reality leads me to add that by far the most enlightening

discussion of models I have found is in Harré's excellent introduction to

the philosophy of science, "The principles of scientific thinking" [37]. He
writes from a realist point of view which eschews simplifications and
attempts to present a theory of science based on the actual complexity of
scientific theory and practice; he regards the alternative traditions of
conventialism and positivism as vitiated by the attempt to force the
description of scientific intuition and rationality into the deductivist
mould. Model building becomes an essential step in the construction of a
theory. I shall not attempt to summarize the argument of his second
chapter, which demands careful and considered reading, but it may be useful
to mention one or two of the distinctions he makes. He starts with the
notion of a sentential model in which one set of sentences T is a model of
(or with respect to) another set of sentences S if for each statement t
of T there is a corresponding statement s of S such that s is true
whenever t is acceptable and t is unacceptable whenever s is false.
If T and S are descriptions of two systems M and N and T is a
sentential model of S, then M is an iconic model of N. He recognizes
that in mathematics the word is used in both ways: model theory is clearly
a sentential model within mathematical logic, but we often conceive sets of
objects, real or imaginary, which are described mathematically. The latter
is an iconic model and the equations a sentential model of the sentences
describing the set of objects. Harré goes on to distinguish between the
subject and source of a model. The former is whatever the model is a model
of, the latter what it is based on; for example elementary kinetic theory
gives models of a gas (subject) based on the mechanics of particles (source).
Homeomorphs are models in which the source and subject are the same as in a
mechanical scale model. When source and subject are not the same, as with
the English tubes and beads that amazed Duhem so much (cf. Sec. 2.1), Harré

6

speaks of paramorphs. He goes on to discuss the taxonomy of models and to show how they are incorporated into the construction of theories first by the creation of a paramorph and then by supposing that it provides a hypothetical mechanism. This process evokes existential hypotheses and raises such questions as the degree of abstraction that can be tolerated leading into a full-scale discussion of the formation of scientific theories. Clearly mathematical modelling in the sense in which we are here discussing it is a small part of this much larger design.

1.2   Relations between models with respect to origins.

It seems well to use the term model for any set of equations that under certain conditions and for a certain purpose provide an adequate description of a physical system. But, if we do this, we must distinguish the kinds of relationships that can obtain between different models of the same process. (This approach seems more useful than to talk of models and sub-models, since the relations are more varied and mixed than can be compasses by this nomenclature). It is of the first moment to recognize that models do not exist in isolation and that, though they may at times be considered in their own terms, models are never fully understood except in relation to other members of the family to which they belong.

One type of relationship can be seen in the packed bed example, the full details of which are given in Appendix A. The physical system is that of a cylindrical tube packed with spherical particles and our purpose is to model the longitudinal dispersion phenomenon. By this we mean that if a sharp pulse of some tracer is put into the stream flowing through a packed bed it emerges as a broad peak at the far end of the bed, showing that some molecules of the tracer move faster through the bed than others and that the sharp peak of tracer is dispersed.

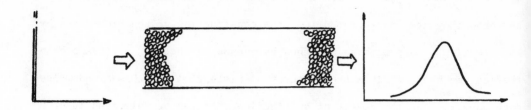

This is the physical system P and it is amenable to modelling in various
ways. The most obvious one is to write down the equation of continuity, the
Navier-Stokes equation and the diffusion equation with their several boundary
conditions (eqns. A1-7). This model, which we will call Π_1, is admirably
complete and founded on the fewest and most impeccable assumptions, but, for
two reasons, it is not a very useful model. In the first place, if the
actual geometry of a given packed bed could be used the results would be
peculiar to that bed, making it a good simulation but a bad model in the
senses of Smith. Secondly, even if the geometry were standardized (and this
presents its own difficulties) to, say, a cubic array of spheres the
resulting equations would present ferocious difficulties to computation and
the model would probably remain barren of results. If the models with
standardized and peculiar geometries are denoted by Π_1' and Π_1''
respectively they are clearly distinguishable but very closely related--in
fact almost "non-identical twins".

The second way in which we might try to model P is to say that the same
sort of dispersion is experienced in a much simpler system, namely that of
plug or uniform flow through a tube with a longitudinal diffusion
coefficient. If we call this modified prototype P_2 we can easily derive a
partial differential equation Π_2, which is much simpler than those of Π_1
(see eqns. A8-12). There is no immediate connection between Π_1 and Π_2
though we can imagine that some sort of averaging of the Navier-Stokes

equations over the cross-section of the bed would lead to the plug flow approximation.

On the other hand, we might make something of the fact that in a packed bed the space between particles makes a natural cavity whilst the interstices narrow where the particles touch and the fluid can be thought of as jetting through into the cavity space. In this rather crude sense the packed bed as a sequence of little stirred tanks gives us a modified prototype, say P_3, which can be modelled. To avoid Suppes' criticism, we do not say that P_3 is a model of P, though recognizing that it is popularly called the "cell model" of the packed bed. The model of P_3 (and therefore of P) consists N ordinary differential equations for the time-varying concentrations in the N stirred tanks of P_3. This will be denoted by Π_3 and the equations are numbered A13-16. There is no obvious connection between Π_1 and Π_3 or between Π_2 and Π_3.

A fourth way of modelling the system P would be to regard the system as a stochastic one in which a tracer molecule had at each step in time the options of either moving forward with the stream or of being caught in an eddy and remaining essentially in the same place. This modification of the prototype, say P_4, leads to Π_4 and the equations A25-26. Again there is no immediate or obvious connection between Π_4 and the preceding models. The relationship of these models is expressed in the diagram.

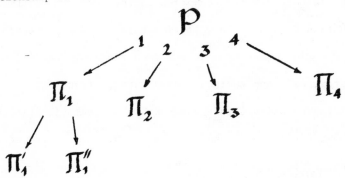

The models $\Pi_1, \ldots \Pi_4$ are best described as cognate models since they appear to be siblings of the same parent system.

A rather different relationship obtains between the models $\Sigma_1 \ldots, \Sigma_6$ of the stirred tank S described in Appendix C. In these Σ_1 is the full set of ordinary and partial differential equations obtained by making mass balances for each of the S species and energy balances on the contents, the wall, and the cooling system of the reactor (see pp. 152-164 for details and particularly pp. 153 and 154 for the hypotheses). This gives S+2 ordinary differential equations and a parabolic partial differential equation. This system, Σ_1, (eqns. C1-6) is again of considerable complexity, but less difficult of calculation than P_1. Σ_2 is the steady state version of these equations obtained simply by deleting all time derivatives and with them the initial conditions. It thus consists of non-differential equations coupled to an elliptic differential equation. If the assumption is made that the wall of the reactor is thin, (hypothesis H_6) the elliptic equation can be solved quite easily and Σ_3 then consists entirely of algebraic equations, C9 and 10. (We will call equations that are not differential equations 'algebraic' even though they may contain transcendental functions).

To reach the model Σ_4 we return to the full transient model Σ_1 and assume that the conductivity of the wall is very high (hypothesis H_7). Then the parabolic partial differential equation can be replaced by an ordinary differential equation for the mean wall temperature and Σ_4 consists wholly of ordinary differential equations one for each reacting species and one for each of the reactor, wall and coolant temperatures. (eqns. C1, 2, 6, 11) If H_6 and H_8, the hypotheses that assert that the wall is thin and of negligible heat capacity, are imposed instead we have one fewer equation and the model Σ_5 (eqns. C1, 13, 14).

10

The final model, Σ_6, is written for a special case of some importance, both historically and logically. It is a case that has dominated the development of stirred tank reactor analysis, as seen in the papers of van Heerden [253], Amundson and Bilous [258], Amundson and Aris [5] and Uppal, Ray and Poore [183,184]. Logically it can be defended as the simplest case in which the essentially nonlinearity of the nonisothermal behavior comes to light. Thus the added hypotheses of instantaneous cooling action (H_9) and restriction to a single irreversible, first-order reaction (H_{10}) allow the system to be immediately reduced to a pair of ordinary differential equations (C15 and 16).

The relationship of those models is represented by the diagram:

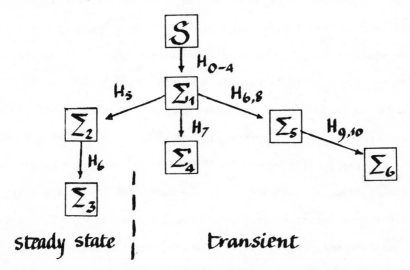

It makes more sense in this case to speak of $\Sigma_2, \ldots \Sigma_6$ as derived from (rather than cognate with) Σ_1 since no modification of S is involved and the later models can be obtained from Σ_1 by letting certain parameters take on limiting values. Steady-state models are worth singling out as of particular importance. They can be obtained formally as a limiting case introducing an artificial parameter multiplying each time derivative and letting this go to zero. But this should be distinguished from a

11

pseudo-steady state hypothesis such as H_9 in which a physical parameter is very small and is allowed to take its limiting value of zero. Thus we have a clone of models in which Σ_2, Σ_4 and Σ_5 are immediately derived from Σ_1 while Σ_3 and Σ_6 are derived from Σ_2 and Σ_5.

The reader is left to justify the diagram given on p. 149 for the relationship of the models of the chromatograph given in Appendix B. They are not predominantly cognate, as are the Π_i, or derived, like the Σ_j, but appear to be a mixture of both. No new relationships seem to be introduced however.

The model of a model represents a relationship which is a little different from that of cognate models or from the idea of a derived model. This kind of modelling arises when the first model is so complicated, either in the form of the equations or the number of parameters, that it seems that better insight can be gained by quite drastic simplification. Such is the case when Burger's equation $u_t + uu_x = \nu u_{xx}$ is used to get preliminary insight into the nature of turbulence. It is not claimed that the physical system corresponds exactly to this equation, though it may be that an artificial one could be constructed. But the model of the model has its validity in so far as it extracts some important feature of the first model with a form in which it can be analyzed more easily. The relationship is more like:

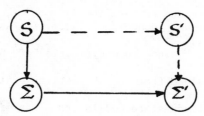

where Σ is the first model and Σ' the model of it. The relationships indicated by the dotted line may or may not exist and even when they exist they may or may not be worth attending to.

Another example is the Lorentz equations

$$\dot{x} = -\sigma(x-y)$$

$$\dot{y} = -xz + rx - y$$

$$\dot{z} = xy - bz$$

which have a tenuous connection with meteorology, whose equations are well developed but vastly complicated; they are perhaps best regarded as a model of the meteorological equations. How profitable, and indeed fascinating, the study of them may be is to be seen from Lorentz' [119] and Marsden and McCracken's [128] treatments of them.

In some cases the difficulties with Σ, the model of the physical system S, lead the investigator to consider a simplified system, S', and construct a model, Σ', of it. In this case the connection between Σ and Σ' may not be of interest and the situation is

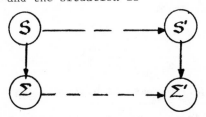

Such a case is the famous "Brusselator" where the unlikely reaction mechanism $A \to X$, $B+X \to Y+D$, $2X+Y \to 3X$, $X \to C$ was inspired by the Belousov-Zhabotinsky reaction. It has much intrinsic interest and served to bring to light some important phenomena (see e.g. Nicolis and Portnow [135] Lavenda, Nicolis and Herschkowitz-Kaufmann [109] or the introductory summary in [14]. But it has little to do with the Belousov-Zhabotinsky reaction itself, the "Oregonator" and other mechanisms being much more direct attempts to give precursors of a model for this. The reader should refer to the excellent little monograph of Tyson [182] and the references given there, particularly the papers of Kopell and Howard, Murray, Othmer and Winfree [see cf. 219,254].

13

Another example stimulated by the need to reduce both the number of parameters and the number of equations in Aris and Schruben's [17] simplification of Amundson and Liu's [6] packed bed equations. Defining the heights of transfer units for mass and heat between the particles and intersticial fluid as H_g and H_h, respectively, Amundson and Liu had written

$$\frac{\partial p}{\partial x} + \frac{1}{u}\frac{\partial p}{\partial \theta} = \frac{p_p - p}{H_g} \tag{1.1}$$

$$\frac{\partial t}{\partial x} + \frac{1}{u}\frac{\partial t}{\partial \theta} = \frac{t_p - t}{H_h} \tag{1.2}$$

where p = partial pressure of the reactant in the fluid,

p_p = partial pressure in the particle where reaction takes place.

t = temperature of fluid,

t_p = temperature of particle,

x = distance from inlet

θ = time

These equations were derived from mass and heat balances in the fluid and similar handling of the particles, assuming a first order reaction rate $k(t_p)p_p$, gave

$$L_g \frac{\partial p_p}{\partial \theta} = p - p_p - M_g k(t_p)p_p \tag{1.3}$$

$$L_h \frac{\partial t_p}{\partial \theta} = t - t_p - M_h k(t_p)p_p \tag{1.4}$$

where $L_g, \ldots M_h$ are further combinations of the flow rate, particle size, density of fluid, etc., the details of which need not concern us here. If we make p, p_p dimensionless by dividing by p*, t and t_p by t*, x with x* and θ with θ^* we are left with five parameters $u\theta^*/x^*, H_g/x^*, H_h/x^*, L_g/\theta^*, L_h/\theta^*$ and the additional parameters in $M_g k(t_p)$

14

and $M_h k(t_p) p*/t*$ of which there are three. Of these eight only four can

be eliminated by the ch ice of $x*$, $\Theta*$, $p*$ and $t*$; for example, if

$x* = H_g$, $\Theta* = H_g/u$, $p* = t*/M_h A$, $t* = E/R$ where $k(t_p) = A\ exp-(E/RT_p)$, we

are still left with $L_g u/H_g$, $L_h u/H_g$, H_h/H_g and $M_g A$. In the steady state

the first two disappear, being multipliers of the time derivatives.

If we look at the steady state equations we have

$$H_g \frac{dp}{dx} = p_p - p = -M_g k(t_p) p_p \tag{1.5}$$

$$H_g \frac{dt}{dx} = (H_g/H_h)(t_p - t) = m M_g k(t_p) p_p \tag{1.6}$$

where $m = M_h H_g / M_g H_h$. Thus from the two end terms of the equations

$$H_g \frac{d}{dx} [mp + t] = 0 \tag{1.7}$$

or

$$p = p_e - (t - t_e)/m \tag{1.8}$$

where p_e and t_e are the entrance values. This last equation merely

expresses the fact that the bed is adiabatic. Also from the last two terms

of eqn. (1.6).

$$t_p - t = \frac{M_h p}{M_g} \frac{M_g k(t_p)}{1 + M_g k(t_p)} \tag{1.9}$$

Using eqn. (1.8)

$$Q_1 \equiv \frac{M_g}{M_h} \frac{t_p - t}{p_e + (t_e/m) - (t/m)} = \frac{M_g k(t_p)}{1 + M_g k(t_p)} \equiv Q_2 \tag{1.10}$$

The right hand side of this equation is an S-shaped curve depending only on

the two parameters in $M_g k(t_p)$. The left hand side varies from point to

point in the bed since it depends on t. However it represents a family of

straight lines all passing through the point $(t_e + mp_e, H_g/H_h)$. For any t

at position x, t_p can

15

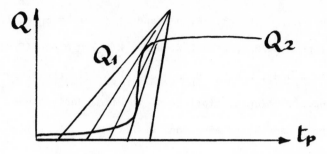

be found by solving eqn. (1.10) for t_p and eqn. (1.5) for p_p. Then with

p given by eqn. (1.8), there is a single equation

$$\frac{dt}{dx} = f(t) \tag{1.11}$$

to solve for t(x) and this can even be done by quadrature. However the

diagram shows that the solution of eqn. (1.11) may be multivalued and this

is the origin of the multiple profiles that are possible in a packed bed.

Eigenburger [59,60] showed that this continuum of steady states is reduced

to a single state if heat conduction between particles is allowed, but the

same phenomenon can be found in isothermal beds with more complicated

kinetics.

Because of the number of parameters there seems no possibility of getting

a comprehensive view of the system, though it is to be noted that the

multiplicity of the solutions of eqn. (1.10) is determined by the study of

the stirred tank in Sec. 5.2, if

$$\alpha = M_g A, \quad \zeta = RM_h p/M_g E, \quad v = Rt/E.$$

This suggests that a simpler model should be constructed that would incor-

porate this feature of multiplicity with the fewest parameters possible.

This was the motivation of Aris and Schruben [17].

Note first that the intermediate intersection can be disregarded since it

is known always to be unstable. This suggests that the sigmoid curve should be replaced by a step function i.e.

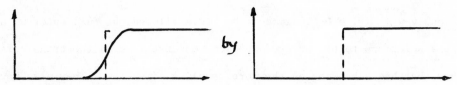

Also it seems desirable to reduce the number of equations to two. This can be done if we suppose that the wall of a tube generates heat at a rate Q by means of an exothermic reaction of order zero which is triggered at a critical temperature T_c. This heat from the wall, whose temperature at position x and time t is denoted by $W(x,t)$ is transferred to the fluid flowing in the tube, whose temperature is $T(x,t)$, and we have the equations

$$c_w a_w \frac{\partial W}{\partial t} = h_p(T-W) + QH(W-T_c)$$

$$c_f a_f \left[\frac{\partial T}{\partial t} + V\frac{\partial T}{\partial x}\right] = h_p(W-T)$$

where c_w, c_f are the heat capacities of wall and fluid,
a_w, a_f are their areas,
p is the inner perimeter of the tube,
h is the heat transfer coefficient,
H is Heaviside's step function,
and V is the velocity of the fluid.

The substitutions

$$v = h_p(T-T_c)/Q, \quad w = h_p(W-T_c)/Q$$

$$\xi = h_p x/V c_f a_f, \quad \tau = h_p t/c_f a_f$$

$$\omega = c_w a_w/c_f a_f$$

give

$$\frac{\partial v}{\partial \tau} + \frac{\partial v}{\partial \xi} = w-v$$

$$\omega \frac{\partial w}{\partial \tau} = v-w + H(w)$$

17

These equations have only one parameter and the solution of the nonlinear equation v–w + H(w) = 0 is immediate. This allowed the full spectrum of possible solutions to be surveyed and some observations on the transients to be made that had a bearing on the more complicated system. They also have an important bearing on the behavior of the monolithic reactor.

1.3 Relations between models with respect to purpose and conditions.

The relationship between models is not only an intrinsic matter of mathematical genealogy but must be viewed also in the perspective of the purpose of the model and the conditions under which it is to be used. For example, the steady state models of the stirred tank, Σ_2 and Σ_3, are quite unfitted to the purposes of control whatever the conditions may be. They are adapted to the purpose of steady state design however. Σ_2 may be demanded by the conditions of thick walls or poor heat conduction but Σ_3 suffices otherwis

This is indicated in part (a) of the figure below, where the boundaries of

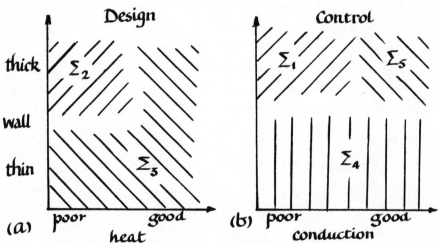

the regions are deliberately vague to indicate that the division between the regions of applicability is not a sharp one. If the purpose is control then Σ_1, the most complicated model, may be demanded by the same conditions that demand Σ_2 for the steady state. Σ_4 is valid if the wall is thin whatever

18

its conductivity and Σ_5 is a simplification made possible by the good conductivity of the wall even when it is thick. Another way of representing
the interaction of conditions and purposes with the type of model is shown
in the next figure. Here the degree of sophistication of the model ranges
from mere algebraic equations at the lowest level to the coupled partial and
ordinary differential equations of Σ_1.

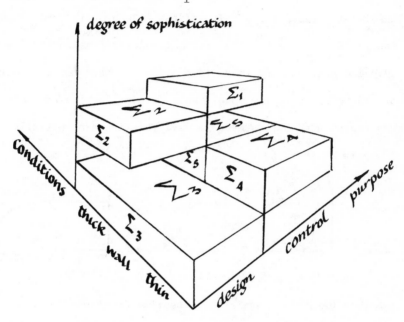

Sometimes the regions of applicability of models can be delineated more
exactly by setting up a certain standard of accuracy. Though this is
admittedly arbitrary it sets the stage for the entrances and exits of the
models. An example is contained in the early work of Gill and his colleagues
on Taylor diffusion [74,75] which may be illustrated by the Δ-models of
Appendix B. The problem is to describe the movement of solute as the solvent
passes in laminar flow through a long tube. The combined influence of
molecular diffusion and convection softens an initially sharp front, for,
thanks to the parabolic flow profile, tracer molecules at the center of the
tube are taken ahead of the pack by the fast central streams but, then

19

finding themselves in a region of low tracer concentration, can diffuse out-wards to the slower streams and so slow down. The mean concentration of tracer is thus diminished in an error-function fashion about a point moving with the mean speed of the stream. In fact this mechanism of dispersion, first elucidated by Taylor in 1953 [174], shows that the higher the diffusio coefficient the smaller the longitudinal dispersion, since lateral diffusion will immediately negate the effects of the flow profile. But when the diffusion is isotropic a high lateral molecular diffusion implies a high longitudinal dispersion and there is a point at which total effect is minimum. The purpose of the model is to account for the advancing wave of solute as represented by its mean value. This is zero initially and in a general way is as shown here:

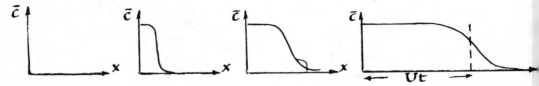

The models, described in detail in Appendix B, are:

Δ_2 the full parabolic partial differential equations to be solved for the concentration c as a function of x, distance from inlet, r, distance from tube axis, and time, t. This solution is averaged over the cross section to give $\bar{c}(x,t)$.

Δ_4 the equation for plug flow at the mean speed of the stream with an equivalent longitudinal diffusion coefficient D_e are solved for the mean concentration directly. D_e is related to D, the molecular diffusion coefficient, a, the tube radius and U, the mean velocity.

Δ_4' $D_e = a^2 U^2 / 48D$

Δ_4'' $D_e = D + A^2 U^2 / 48D$

Δ_4''' $D_e = D$

Δ_5 A pure convection model with no consideration of diffusion

Δ_6 An empirical fit of D_e between the solutions of Δ_2 and Δ_4.

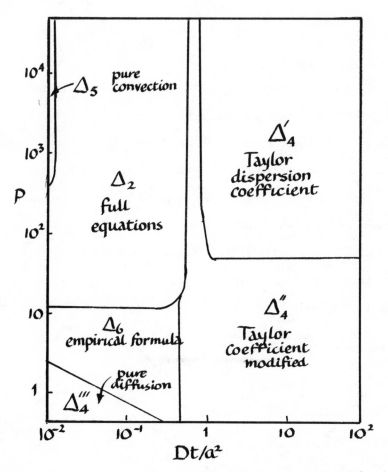

If the purpose is to provide a sufficiently accurate value of $\bar{c}$ for a period of time t we may take one axis to be the dimensionless time $\tau = Dt/a^2$. The conditions are represented by the Peclet number $P = aU/D$. Gill and his colleagues [74] mapped the τ–P space as shown above. Thus, for example, the full equations have to be solved in the region Δ_2, but the Taylor diffusion coefficient $a^2U^2/48D$ in the plug flow model suffices in the region Δ_4'. The arrangement of the Δ_4-regions is reasonable, for the modification Δ_4'' (cf. [10]) can be written $D[1 + P^2/48]$. Thus when P is large we can neglect the 1 in the bracket giving Δ_4', whereas, when P is small, $P^2/48$ is negligible. Δ_6 is an empirical bridge between Δ_4'' and Δ_4'''.

1.4 How should a model be judged?

We shall have more to say of the detailed evaluation of models in the last chapter, but it will be useful to make a few points in a preliminary way. Clearly, once a model has been set up it has a life of its own and its equations have their own intrinsic interest. However most models are indissolubly bound to their origins and cannot be viewed in isolation from their background any more than they can ignore their relatives. The relation is a dynamic one calling for continual interaction if either conceptual progress or actual understanding is to be gained. It may be envisioned in the following way:

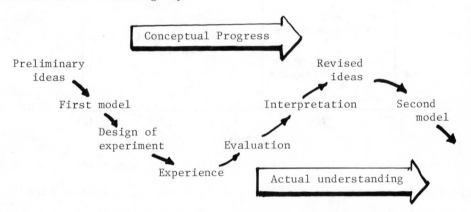

A certain primacy belongs to theory since it is impossible to design a good experiment without some theoretical vision and the more precise the theory the more decisive the experiment can be. If conceptual progress is to go hand in hand with the understanding of an actual situation there must be this intercourse between the system, S and its family of models, Σ, though the distinction that Smith has made between model and simulation is a valid one. It might be represented by a serpentine progress that tends to emphasize one level of the other, i.e.:

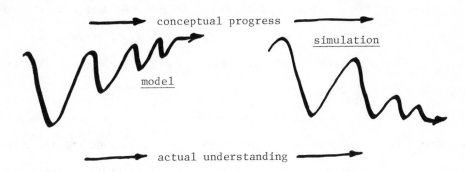

The revision of ideas and development of models is not necessarily in the direction of greater complexity or an increasing number of parameters. Progress may be toward simplification and the reduction of the number of adjustable constants since it is often said that you can fit an elephant with five constants, though Wei [189] has shown that the fit may not be spectacular. While it is certainly convincing if a complicated situation can be represented by the adjustment of very few constants, care must be taken to see that it is not purchased by a less obvious accommodation else-where, as when it was said of a certain theorist that his work required no adjustable constants but completely pliable hypotheses. Nevertheless part of the judgment of a model will lie in whether its constants can be found from independent sources and combined to give a convincing picture in the interactive situation. Thus we have a high degree of confidence in a model of a reactor embodying such complicating features as diffusion within catalyst pellets if the kinetics of the reaction, diffusivities and catalyst properties can all be determined independently and give, in the model, a

23

recognizable behavior. We have less confidence if we adjust the parameters
in the model as a whole, for example by a least square fit of outputs, without any interior understanding. Similarly at the interior level in determining, say, the kinetics, we have much more confidence in a kinetic
expression that is based on an independently confirmed mechanism than in an
overall kinetic expression that lacks this insight. Of course, the
exigencies of a particular situation may force us to be content with less
than the best.

There is however a phenomenon of current interest that raises deeper
questions about our ability to compare the implications of a model with
experience. This is the class of solutions of quite simple equations which
are said to be chaotic in their behavior. The locus classicus in which this
behavior may be observed is the difference equation

$$x_{n+1} = \lambda x_n (1-x_n)$$

--an innocent enough starting point for anyone. If $0 \le \lambda \le 4$ the transformation maps the interval [0,1] into itself. The origin $x = 0$ is always
a fixed point and indeed is the only fixed point until λ exceeds 1 when
$1-\lambda^{-1}$ is also a fixed point. At $\lambda = 1$ the origin becomes unstable but
the new fixed point is stable and remains so until $\lambda = \lambda_2 \equiv 3$. At this
point two stable solutions of period 2 (i.e. for which $x_{n+2} = x_n$) appear
and are stable. But not for long, since when λ reaches $\lambda_4 = 1+\sqrt{6} = 3.45$
these become unstable and spawn four solutions of period 4 which are, at
first, stable. This process of binary fission becomes increasingly frequent
as λ increases and in fact the sequence of points $\lambda_2, \lambda_4, \lambda_8, \ldots$ at
which the 2^n stable cycles of period 2^n appear has a limit point $\lambda_c = 3.57$.
Beyond this point there are a countable infinite number of unstable periodic
orbits and also an infinite number of solutions which are in no sense

24

periodic, that is they are neither periodic nor asymptotic to a periodic solution. At $\lambda = 3.68$ a long periodic solution of odd length appears and by the time $\lambda = \lambda_3 \equiv 3.83$ a cycle of period 3 is seen. At this point cycles of all periods are present some of which are stable. In fact cycles of period 3 bifurcate into those of period 6, which in turn go to those of period 12 and again the values of λ at which these bifurcations take place have a limit point ($\lambda = 3.85$) short of $\lambda = 4$. Indeed Li and Yorke [115,222] have shown that once there is a cycle of period 3 there must be cycles of all periods as well as strictly non-periodic solutions--a situation aptly described as chaotic.

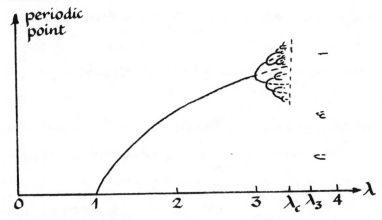

This remarkable situation is not peculiar to the expression $\lambda x(1-x)$ but is generic for all functions with a hump and a parameter such as λ by which it can be 'tuned'. It is described with admirable clarity by May, Oster and Guckenheimer in various places [128,129,130,211]. It is a feature shared by many allegedly simple systems as the Lorentz equations [119] and the examples of Rössler [157] show. In itself it points up the value of even the simplest models for here a simple generic situation has opened our eyes to a new type of behavior of the solution which may well reflect the irregularities experienced in nature. It differs from a random process in the following sense. In the random process the attempt to predict future

25

states is limited by the range of the correlation of the random process, whereas in a chaotic process it is limited by the accuracy with which the initial conditions can be determined. This is the case because arbitrarily near to the initial point of any solution there are infinitely many initial points that will give solutions that ultimately diverge completely from the first solution. But this raises the question of whether the matching of the results of a computation with experience can ever be trusted. Such questions as: does a mismatch constitute an adverse reflection on the model or is it only the result of a failure to find the initial conditions with sufficient accuracy? could a chaotic solution be distinguished from one of a very long period? even if such a distinction could be made would it matter? how can two models ever be compared if their solutions are both chaotic? These and other questions are as yet unanswered but are relevant to the question of evaluating a model.

Different types of model call for different modes of evaluation. Bush and Mosteller [38] present an interesting comparison of eight statistical models of a learning process and in doing so examine various criteria. They reject the likelihood ratio, for example, as an adequate tool for discrimination. Though it is a convenient summary of the fit of the model to the data it is often difficult to compute and obscures the peculiar strengths and weaknesses of a particular model, failing to suggest why the model is inadequate. Moreover, it may be sensitive to uninteresting differences between the model and the experimental set-up. Instead they use a number of different statistics (e.g. trials before first avoidance, trials before last shock, etc.) to judge the match between the data and the learning sequences calculated from each model.

2 The different types of model

"Let us take them in order. The first is the taste,

which is meagre and hollow, but crisp: ...."

C. L. Dodson. The Hunting of the Snark.

Fit 2. St. 16.

This chapter is little more than an annotated list of various types of model
that are in common use. It would be pleasant, but unrealistic, to think that
the author could produce a sequence of sparkling little essays on the his-
torical origins, the current status, the manifold areas of application and
the virtues and vices of each. No doubt this would be the very model of such
a chapter but, for the moment, this rather superficial survey must suffice as
a reminder of the variety of tools that the modeller should have in his bag.

2.1 Verbal models and mechanical analogies.

This type of model, referring to one couched in the language of everyday
discourse rather than in the language of mathematics, is in the strict sense
outwith our present policies, but it should nevertheless be mentioned since
it may well have some of the characteristics of a mathematical model. Such
models might be called "soft" models without being pejorative for they are
useful without being "hard" in the sense of having quantitative predictions
or sharply defined concepts. Toffler [88, p. 274] points out that in the
social context it is virtually impossible to make absolutely explicit all
the assumptions of a hard model and that models can thus be "implicitly soft."
On the other hand there are "explicitly soft" models and Toffler points to

the novelists who present verbal models of society often with great precision; e.g., McCarthy of the internal politics of a small United States
college in "The Groves of Academe" or Snow's "The Masters" with its model of
a Cambridge college. (The ambiguity--and richness--of verbal models is
evident when one considers the totally different overtones of the one word
"college.") He is in line with Maynard Smith's use of the word when he
writes: "Good mathematical models don't 'predict' in the colloquial sense
of the word. But they can broaden our understanding of the potential con-
sequences of our decisions..." In a very different context from the Venice
seminar of economists, sociologists and futurists to whom Toffler spoke,
Oster and Guckenheimer have reported [128, p. 328] "that many ecologists
seriously question whether mathematics can play any useful role in biology.
Some claim that there has not been a single fundamental advance in biology
attributable to mathematical theory. Where complex systems are concerned,
they assert that the appropriate language is English, not mathematical."
These are reservations worth bearing in mind for mathematical modelling is
not without a certain danger of narcissism (cf. [224]).

   The use of mechanical analogies, verbally described, is not uncommon in
mathematical modelling. Rheologists will, for example, often talk about
springs and dashpots even when the equations they obtain could have been
derived directly from more abstract hypotheses. When they draw further
pictures of the coiling and uncoiling of polymer chains they may indeed be
making mechanical hypotheses. But analogies also have their place. How much
of a place is to be given to them is partly a matter of taste, and there is
a famous passage in Duhem [56] contrasting like French and English in their
approach to the physics of his day. "The employment of similar mechanical
models...is a regular feature of the English treatises on physics. Here is

a book intended to expound the modern theories of electricity...in it there are nothing but strings which move around pulleys, which roll around drums, which go through pearl beads, which carry weights; and tubes which pump water while others swell and contract; toothed wheels which are geared to one another and engage hooks. We thought we were entering the tranquil and neatly ordered abode of reason, but we find ourselves in a factory."

2.2 Finite models.

The theory of graphs has found many natural applications in the physical and social sciences. A graph is a collection of vertices, V, linked in some way by elements of a collection of edges, E. If u, vϵV are vertices (the terms node and point are also used), then uvϵE denotes the edge connecting u to v. If this has a sense of direction from u to v, the graph is a directed graph or digraph. E is thus a subset of VxV. If vv is allowed as a proper element of E, then the graph is called a loop graph. If more than one edge may connect two vertices, the term multiple graph or multigraph is often used. The problems of connectivity and decomposition, or characterization and the topography of graphs arise immediately and there is a rich theory on which to draw. The most obvious applications are to networks of all sorts, electrical, mechanical, transportation, job assignment and scheduling, industrial inter-dependence and the planning of experiments. Perhaps the most famous "application" is the four-colour problem, the many, unsuccessful attempts to solve which have contributed greatly to the advancement of graph theory. But the structure of many different areas, social, physical and intellectual, can be illuminated by its methods; see, for example [85]. It has even been applied to the structure of Mozart's "Cosi fan Tutte" [83]. It has a vast literature of which only a sample can be mentioned: [24,27,28,67,84,85,140].

The theory of games is another finite model that has found wide application and indeed was developed by von Neumann and Morgenstern in the context of economics. In its simplest form two players P and Q each select one of their set of options $(p_1, \ldots p_m)$ for P and $(q_1, \ldots q_m)$ for Q, the payoff from P to Q if the choices are p_i and q_j being an amount a_{ij}. The question is whether there is an optimal way for each to play. A pure strategy for p is the choice of one p_i, a mixed strategy, σ, is the choice of a set of numbers s_i, $i=1, \ldots m$, $\Sigma s_i = 1$, which can be regarded as the proportion of times in a long run that p_i will be chosen. A similar definition applies for t_j, a mixed strategy τ for Q. The expected payoff $p(\sigma, \tau) = \Sigma\Sigma s_i a_{ij} t_j$ and the game has a value if there exist two strategies $\bar{\sigma}$ and $\bar{\tau}$ such that for all σ and τ

$$p(\bar{\sigma}, \tau) \geq p(\bar{\sigma}, \bar{\tau}) \geq p(\sigma, \bar{\tau}).$$

The fundamental theorem shows that a two-person game has a value under the condition that it is zero-sum, i.e., the payoff to Q of an amount a is the same as a payoff to P of -a. There are many natural applications in the social and physical sciences. Some problems in control theory can be regarded as games against Nature. There are also many extensions, as to non-zero-sum and multiperson games and to differential games in which the action and condition of the players develop continuously in time. See, for example, [69,99,121,149,150,191].

Game theory is intimately connected with linear programming, the problem of determining the set of non-negative x_j, $j=1,2,\ldots N$, that satisfy

$$\sum_{j=1}^{N} a_{ij} x_j \leq b_i, \quad i=1,2,\ldots,M$$

and maximize

$$Z = \sum_{j=1}^{M} c_j x_j.$$

To which there is a dual problem, that of minimizing

$$W = \sum_{i=1}^{M} b_i y_i$$

subject to

$$\sum_{i=1}^{M} y_i a_{ij} \geq c_j, \quad j=1,2,\ldots N.$$

Many systems have been modelled in this form ranging from huge input-output models of the economy to modest problems of blending. Dantzig, the principle architect of the subject, has given a splendid exposition of it [52] and its extensions in a book from which one can learn much about modelling in general.

Finite automata have been used variously as models, not only in computer science, the house in which they were born, but also in control theory, linguistics, psychology, and biology. See [8,9,62,78,168,180]. Basically, the automaton is the computer reduced to its simplest elements, an input/ output tape on which symbols from a finite alphabet are read or written and a set of internal states. The computation is done by a set of instructions that modify the internal state and either move the tape or modify the symbol under the head. If the alphabet is $A = \{a_j; j=1,\ldots M\}$ and the states are a set $S = \{S_k; k=1,2,\ldots N\}$ the instruction can be of three forms: $(a_j, S_k) \rightarrow (a_p, S_q)$ says that when the machine is in state S_k and reads a_j it replaces the symbol by a_p and changes its state to S_q; $(a_j, S_k) \rightarrow (1, S_q)$ changes the state to S_q but leaves a_j unchanged merely moving the tape one place to the left; $(a_j, S_k) \rightarrow (r, S_q)$ does the same except that movement is to the right. The machine is deterministic if (a_j, S_k) has a unique consequence. A computation is a series of such steps which either terminates or goes into a repetitive sequence. Usually there is a distinguished initial state and the input tape, which is finite, starts in the left-most position. When it stops, the state of the tape may be regarded as the output. The Turing machine, as this automaton is called, thus gives a mapping from the input tape to the output tape. It can be generalized to have several reading and writing heads, but it is a matter of convenience rather than necessity for Church's thesis is that all devices that formalize the notion of computability are equivalent. The finite automaton is a Turing machine that only reads and only moves the tape in one direction, say to the left; its instructions can therefore be written $(a_j, S_k) \rightarrow S_q$ since the 1 can be taken for granted. It is not hard to see how such a device has possibilities for representing the learning process or the formal aspects of language and these and other developments are to be found in the references given.

2.3 Fuzzy subsets.

An important class of mathematical models was introduced by Zadeh [195] in 1965 when he defined a fuzzy set or subset. The usual definition of a subset A of U can be formalized in terms of the characteristic function of the subset $\chi(x)$ such that for any $x \epsilon U$, $\chi(x)=1$ if $x \epsilon A$ and $\chi(x)=0$ if x does not belong to A. Such definiteness is all very well in its place but there are clearly many situations whose intrinsic ambiguity and vagueness is ill-served by such a black-or-white attitude. The concept of fuzzy subset replaces the characteristic function $\chi(x)$ with values in the set $\{0,1\}$ by a membership function $\mu(x)$ with values in a membership set M. Usually M is a totally ordered set (very often the closed interval $(0,1)$ is taken) and $\mu(x)$ is the degree of membership of x in A. For example, if U is the real line and $M=[0,1]$, the fuzzy subset of "small numbers" might be defined with $\mu(x)=(1+|x|)^{-1}$ or of "really small numbers" by $\mu(x)=(1+|x|)^{-50}$.

The usual operation with sets can be defined suitable for fuzzy subsets. For example, if A and B are fuzzy subsets of U with the same membership set, A is included in B is $\mu_A(x) \leq \mu_B(x)$ for all $x \epsilon U$. The union of two fuzzy

32

subsets has a membership function $\mu(x) = \text{Max}(\mu_Z(x), \mu_B(x))$. For example, if
M and U are both [0,1], these ideas can be shown graphically.

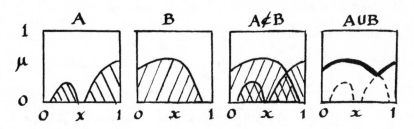

In the same spirit graphs, relations and equivalence can be generalized to
fuzzy graphs, fuzzy relations and similitude. Binary operations on two
fuzzy subsets can be defined leading to monoids and groupoids. For example,
if $U = R_+$ and I_n is the fuzzy subset with membership function
$\mu_n(x) = \lambda^n x^{n-1} e^{-\lambda x}/(n-1)!$ then we can define the composition of two of these
subsets $I_m * I_n$ as the fuzzy set with membership function

$$\int_0^x \mu_m(x-t)\ \mu_n(t)\ dt$$

Then $I_m * I_n = I_{m+n}$ and, if we add I_0 with $\mu_0(x) = \delta(x)$, we have a monoid
$I_0, I_1 \ldots$ of fuzzy subsets which is isomorphic to the natural numbers. The
I_n are called the exponential fuzzy integers.

But the generalization have gone far beyond this and there are fuzzy
categories, topological spaces, logics, algorithms, automata, languages, and
environments. Some of the most important applications concern decision-
making in a fuzzy environment; see, for example, Bellman and Zadeh [26]. The
book of Kaufmann [104] is a lucid introduction; Negoita and Ralescu [134]
are very good but poorly translated; see also Bellman and Giertz [25]
and Zadeh, Fu, Tanaka and Shimura [192].

2.4 Statistical models.

In his analysis of a system S, Bury [39] lumps together the process and its measuring devices as a black-box out of which comes perceived data. The data generated by the process can be corrputed by either systematic or random error in the process of being perceived. The aim of statistical analysis is thus to construct a statistical model on the basis of the available output and hence form conclusions about the underlying phenomena. This is to view the measurements as realization of a random variable X. The statistical model of the process is thus the probability density function (sometimes called the probability mass function when it associates non-zero probabilitie with points) or its integral, the cumulative distribution function, for the random variable. The statistical tools needed to make the desired inferences are an understanding of sampling theory and order statistics and the various qualities of inference--consistency, bias, minimum variance, efficiency, etc.--and the several estimation and confidence tests. There is a range of distributions (Gaussian or normal, log-normal, gamma, beta, binomial, Poisson, Weibull), each with its own virtues and properties.

Of the vast array of books on probability and statistics it is almost impertinent to single out one or two. Kendall and Stuart's magisterial volumes may still be given pride of place [106], but Feller [64], Lindley [117] and Parzen [143] might be mentioned. There is, of course, a huge literature on the philosophical issues in "the matter of chance" [132].

2.5 Difference and differential equations.

Differential equations will so dominate the rest of these notes that it would be somewhat gratuitous to do more than mention them here. They are by far the commonest type of model in the physical sciences, ordinary differential equations playing the same role for lumped systems as partials

34

do for distributed. Whole tracts of control theory, for example, can be
cast in the form

$$\dot{x} = f(x,u,t), \quad y = g(x,u,t),$$
(2.1)

where $x = x(t)$ is a vector of state variables, y a vector of observations
and u a vector of control variables. Feedback control seeks a function
$u = h(y)$ which will attain certain goals; optimal control theory is concerned with finding the control u which will maximize or minimize some
functional of the path or function of the final state. There are key
questions of observability and controllability in such models which lie at
the root of the question of whether optimal or feedback controllers exist.
Naturally these questions are answered most completely for linear systems.
Similar questions pertain to partial differential equations where they are
of course much more difficult to answer.

Difference equations are appropriate when the dependent variable is
discrete. An example has already been given of the logistical equation
$x_{n+1} = \lambda x_n (1-x_n)$. Sampled continuous systems are also an avenue to
different equations. Their theory parallels that of differential equations
in many ways though the example just given does suggest that their behavior
can become bizarre much earlier. The difference-differential equation

$$\dot{x}(t) = f(x(t), x(t-1))$$
(2.2)

is an example of a functional differential equation. Whereas eqn. (2.1)
requires initial values $x(0)$ to be specified if a particular solution is to
be determined, eqn. (2.2) requires all values $x(t)$ over a unit initial
interval, say $-1 \leq t \leq 0$. In that the differential equation has to be dis-
cretized when its solution is to be computed on a digital machine it is
clear that the difference equation is more important than one would first
suppose.

The solution of $\dot{x} = f(x,t)$ also satisfies

$$x(t) = x(0) + \int_0^t f(x(t'),t')dt' \qquad (2.3)$$

which is an integral equation. It can be regarded as a nonlinear integral equation and if X is a suitable defined class of functions the right hand side of eqn. 2.3 is a nonlinear transformation from X into X. If this transformation has a fixed point it provides a solution of the integral equation. Integral equations and indeed integro-differential equations are used as models, though we shall not have occasion to refer to them.

2.6 Stochastic models.

The integral formulation (2.3) of the differential equation is often used as a starting point for the study of differential equations with random elements. Thus if α is a random variable, i.e. known only through its probability distribution function $F(a) = Pr\{\alpha < a\}$ or its density $f(a)da = Pr\{\alpha \epsilon(a, a+da)\}$, a generalization of eqn. (2.1) is

$$x(t;\alpha) = f(x(t;\alpha),t;\alpha) + g(x(t;\alpha), t)w(t;\alpha) \qquad (2.4)$$

where w is a random process often taken to be white noise. The initial state is also random $x(0;\alpha) = x_o(\alpha)$ but we have a formal analogue of eqn. (2.3) in

$$x(t;\alpha) = x_o(\alpha) + \int_0^t f(x(t';\alpha), t';\alpha)dt + \int_0^t g(x(t';\alpha),t')dW(t';\alpha) \qquad (2.5)$$

where W is the Wiener process from which w is derived. Unfortunately the second integral almost surely does not exist in the ordinary Riemann-Stieltjes sense and it requires special interpretation (see [100],[126] etc.) Nevertheless it is clear that in such stochastic differential equations we have a class of model of the utmost importance. Nor can we fail to expect some striking new results. Take for example the very simplest of cases, the equation $\dot{x} + \alpha x = 0$, $x_o = 1$ where α is uniformly distributed over the

36

interval $(\beta-\gamma, \beta+\gamma)$. The expected value of α is β and with this expected value for α the solution of the equation is $x(t) = \exp-\beta t$. But $x(t;\alpha) = \exp-\alpha t$ and the expected value of this is $E(x(t;\alpha)) = (\exp-\beta t)[(\sinh\gamma t)/\gamma t]$ which behaves quite differently as time goes on. Seldom can the equation be solved explicitly and one has recourse to moments or equations for the transition probability. Thus if $F(x,t|x_o,t_o)=\Pr\{x(t;\alpha)<x|x(t_o;\alpha)=x_o\}$ and $f = \partial F/\partial x$ and $\dot{x} = m(x,t) + \sigma(x,t)w(t,\alpha)$ then F satisfies a Kolmogoroff or backwards diffusion equation and f the Fokker-Planck equation

$$\frac{\partial f}{\partial t} = \frac{1}{2} \frac{\partial^2}{\partial x^2} (\sigma^2 f) - \frac{\partial}{\partial x} (mf)$$

with a boundary condition from the probability density function of the initial state.

The origins of the theory of stochastic processes lie in the study of Brownian motion and other random walk problems, though Markov, one of the key figures in the development, had interests in linguistic problems also. We have mentioned some elementary stochastic models in Appendices A and B. A key notion of wide applicability is that though only known probabilistically the state at time $t+1$ only depends on the state at time t. This property, often called the Markov property, means that the transition probability between state i at time t and stage j at time $t+1$, $p_{ij}(t)$, is all that need be known. If it is independent of t the process is stationary. The theory develops from random walk problems to Markov chains with discrete time and a finite number of states, to processes with discrete states in continuous time and so to those with continuous state space and continuous time, i.e. stochastic differential equations. Non-Markovian processes can also be considered though with more difficulty. Time series, the outputs of stochastic processes, are also studied for their own sake and prediction and

37

filtering theory plays a key role in many applications. It is easy to see that stochastic models are appropriate to a wide range of situations from learning theory or the mobility of the work force to gunnery and ecology. An excellent bibliography up to 1959 has been edited by Wold [192] and there is an extraordinary range of books on the subject of which a few are: [22, 44,48,100,103,126,139,143,154,166,193,227]. For a graphic presentation of three stochastic processes the introduction to [192] makes interesting reading.

3 How to formulate a model

"'You may seek it with thimbles--and seek it with care;

You may hunt it with forks and hope;

You may threaten its life with a railway share;

You may charm it with smiles and soap--'"

C. L. Dodgson. The Hunting of the Snark.

Fit. 3, St. 8.

Comparatively little needs to be said on this score now that we have reviewed the types of model that are available for the formulation is nothing more than a rational accounting for the various factors that enter the picture in accordance with the hypotheses that have been laid down.

3.1 Laws and conservation principles.

The formulation of the equations of a model is usually a matter of expressing the physical laws or conservation principles in appropriate symbols. This can often be written down as a prescription as, for example, in particle dynamics [107] where, if m is the mass of the particle and $\underset{\sim}{r}$ its position with respect to an inertial frame, $m\ddot{\underset{\sim}{r}}$ is calculated and set equal to $\underset{\sim}{F}$, the resultant of all forces acting on the particle. This is a second order equation and, this being recognized, it will clearly be necessary to specify the initial position, $\underset{\sim}{r}_o$, and velocity, $\dot{\underset{\sim}{r}}_o$, before the model is complete. For a single rigid body whose center of mass is at $\underset{\sim}{s}$ in an inertial frame, $m\ddot{\underset{\sim}{s}}$ is equated to the resultant of the forces and $\dot{\underset{\sim}{H}}_Q$, the rate of change of the vector of moments of momentum, to the moment of these forces about Q, provided that Q is fixed or the center of mass, $\underset{\sim}{s}$.

Except in relativistic contexts when the interconvertibility of mass and energy is at issue, the conservation principles invoked for physical problems are usually those of mass, momentum or energy. We can also do number counts, as, for example, in a population model which may be used to illustrate the basic kind of balance that is involved. Let $n(a,t)da$ be the number of individuals in the age bracket $(a,a+da)$, then we can compute the change in this number during the time interval $(t,t+dt)$. By definition of n this is $\{n(a,t+dt)-n(a,t)\}da$ and by a balance over the age bracket this is the number of individuals who 'age' into the bracket, i.e., $n(a-dt,t)dt$, minus the number who age out of it, i.e., $n(a,t)dt$, minus those that die $\Theta(a,t)dadt$. Thus dividing by dadt

$$\frac{n(a,t+dt)-n(a,t)}{dt} = \frac{n(a-dt,t)-n(a,t)}{da} - \Theta(a,t).$$

We then recognize that age and time run simultaneously so that da=dt. Letting the common value of this increment go to zero we have

$$\frac{\partial n}{\partial t} + \frac{\partial n}{\partial a} + \Theta = 0. \tag{3.1}$$

A more sophisticated version of this basic balance starts by recognizing the simultaneous flow of age and time. This implies that $n(a,t)$ is the flux of numbers that age across a at any time t. Thus, for any interval of ages $b<a<c$ (not necessarily small), $\int_b^c n(a,t)da$ is the total number in that interval $n(b,t)-n(c,t)= -\int_b^c(\partial n/\partial a)da$ is the net flux into the interval and $\int_b^c \Theta(a,t)da$ is the total loss by death. Thus

$$\frac{d}{dt} \int_b^c n(a,t)da = \int_b^c \frac{\partial n}{\partial t} da = -\int_b^c \frac{\partial n}{\partial a} da - \int_b^c \Theta\, da$$

or

$$\int_b^c \left[\frac{\partial n}{\partial t} + \frac{\partial n}{\partial a} + \Theta \right] da = 0.$$

40

But if the integrand is continuous it must vanish everywhere, for, suppose it were positive at a_o, $b \leq a_o \leq c$, then it would be positive in some interval about a_o. But then b and c could both be taken in this interval and the integral could not be zero. The recognition of $n(a,t)$ as a flux across the age a makes it easy to write a boundary condition since $n(0,t)$ is the rate of total births. Thus, if $\gamma(a,t)da$ is the number of births per unit time to individuals in the age bracket (a,a+da) at time t,

$$n(0,t) = \int_0^\infty \gamma(a,t)n(a,t)da. \tag{3.2}$$

This process can be stated rather generally as follows. In a discrete element we can let, F be the net flux of the entity into the element, G its rate of generation there and H the total amount of it which is present. Then F, G and H are functions of time and satisfy

$$F + G = \frac{dH}{dt} \tag{3.3}$$

If we are dealing with a continuum then these quantities must be defined as densities. Thus we let the vector $\underset{\sim}{f}$ denote a flux which is defined such that the flux across an element of area dS in the direction of its normal n is $\underset{\sim}{f} \cdot \underset{\sim}{n}$ dS. Similarly the generation must be defined as a rate per unit volume, so that in a volume element it is gdV, and H becomes a concentration h. Then if Ω is an arbitrary, simply connected region of the continuum with a piece-wise smooth surface $\partial\Omega$ whose outward normal is denoted by $\underset{\sim}{n}$, we have

$$-\int\int_{\partial\Omega} \underset{\sim}{f} \cdot \underset{\sim}{n} \, dS + \int\int\int_\Omega gdV = \frac{\partial}{\partial t} \int\int\int_\Omega hdV$$

In this equation we use the fact that Ω is fixed to interchange the order of integration and differentiation and use Green's theorem on the surface integral. Then all terms can be brought to one side of the equation and we have

41

$$\iiint_\Omega \left[\frac{\partial h}{\partial t} + \nabla \cdot \underset{\sim}{f} - g\right] dV = 0.$$

We must now make the hypothesis that $\underset{\sim}{f}$, g and h are sufficiently continuous that the integrand is continuous and then, since the region Ω is completely arbitrary,

$$-\nabla \cdot \underset{\sim}{f} + g = \frac{\partial h}{\partial t}. \tag{3.4}$$

If the volume is a material volume $\Omega(t)$ moving in a continuum where the velocity field is $\underset{\sim}{v} = \underset{\sim}{v}(\underset{\sim}{x},t)$, then we need Reynolds' theorem for the interchange of differentiation with respect to time and integration. This is

$$\frac{d}{dt} \iiint_\Omega h dV = \iiint_\Omega \left[\frac{\partial h}{\partial t} + v\nabla \cdot h\right] dV.$$

The fact that the flux through a surface element can always be expressed as $\underset{\sim}{f} \cdot \underset{\sim}{n}$ dS is the conclusion of an interesting type of argument that is sometimes useful in other contexts. The figure shows a particular form of elemen namely a tetrahedron of volume dV and with three sides perpendicular to the axes On_1, On_2, On_3 and having dS as the area of its slanting face. Then by definition of the direction cosine the face perpendicular to On_1 has area n_1dS. Let f_i be the flux in the direction On_1 and f the flux over the slant face. Then a balance can be struck over the tetrahedron for

$$F = (f_1n_1 + f_2n_2 + f_3n_3 - f)dS, \quad G = gdV, \quad H = hdV$$

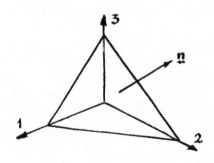

and the equation $F + G = \dot{H}$. But if the volume is allowed to shrink in size, whilst keeping its proportions, dS will decrease as the square of the size but dV as the cube. It follows that G and H become negligible in comparison with F and hence in the limit F = 0. Thus

$$f = f_1 n_1 + f_2 n_2 + f_3 n_3 = \underset{\sim}{f} \cdot \underset{\sim}{n} \qquad\qquad (3.5)$$

if f is the vector with components (f_1, f_2, f_3).

A similar argument is used in the formulation of boundary conditions if the element over which the balance is made is an element of surface extended by a distance dh on either size. Then letting dh→0 first reduces

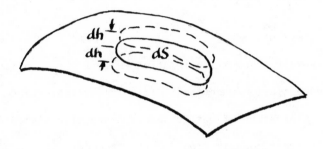

the volume to zero. It is another way of saying that in three dimensions a surface has no volume and hence no capacity for a quantity defined per unit volume. Thus in the boundary conditions (C4) and (C5) there is no term with a time derivative since the natural capacity of the surface is zero. If however a quantity is defined 'per unit area' and is therefore a surface concentration it may well show up in a boundary condition.

The same principles apply to a moving discontinuity as can best be seen in one dimension for a kinematic wave moving in the x-direction

$$\frac{\partial f}{\partial x} + \frac{\partial h}{\partial t} = g$$

If a discontinuity moves with velocity u having a flux f_- and concentration h_- to the left and f_+, h_+ to the right then in a time dt the net flux $(f_- - f_+)$ provides the amount that builds up as the front advances, which in (t,t+dt) is $(h_- - h_+)$udt. Thus the speed of the discontinuity is

$$u = \frac{f_+ - f_-}{h_+ - h_-}$$

We observe that there is no term in g here for there can be no contribution from the discontinuity itself which is a set of measure zero. Put another way, we could say that the amount generated would be $gdxdt = gu(dt)^2$ which becomes vanishingly small in comparison with the other terms as $dt \to 0$.

3.2 Constitutive relations.

In formulating a general conservation relation we left the relationship between flux and concentration undefined. This is desirable since the physical laws (of conservation of matter, etc.) are applicable to a variety of substances of different constitutions. It is the duty of the constitutive relation to provide the connection between f, g, and h or F, G and H. Thus for example in the stirred tank model of Appendix C when the heat equation, (C3),

$$\rho_w c_{pw} \frac{\partial T_w}{\partial t} = k_w \nabla^2 T_w$$

is written for the wall, we assert that

$$\underset{\sim}{f} = -k_w \nabla T_w, \quad g=0, \quad h=\rho_w c_{pw} T_w$$

for the entity 'energy' or 'heat'. Thus the constitution of the wall is that it conducts heat according to Fourier's law, $f \propto -\nabla T$, and does not of itself generate heat, $g = 0$. Similarly the species balance on the whole reactor, eqn. C1,

44

$$V \frac{dc_j}{dt} = q_j c_{jf} - qc_j + \alpha_j Vr(c_1, \ldots c_s, T)$$

is of our general form with

$$F = q_j c_{jf} - qc_j, \quad G = \alpha_j Vr(c_1, \ldots c_s, T), \quad H = Vc_j.$$

The expressions for F and H follow from the definitions but that for G asserts two things about the constitution of the system; first, that the reaction rate is a function of all the concentrations and the temperature; second, that there is but one reaction in which the stoichiometric coefficient of A_j is α_j. This is further specialized by the constitutive relation of the first order irreversible reaction in Σ_6 where $\alpha = -1$,

$$r = k(T)c.$$

Under certain circumstances it might be well to distinguish even the generally applied constitutive relations from the physical laws in the system of hypotheses. This has not been done in the appendices where Fourier's, Fick's law or Newton's law of cooling have been lumped with the basic laws in an underlying hypothesis, H_0, but the more specific constitutive relations have been made explicit (e.g. H_2 and H_{10} for the reaction rate in Appendix C).

Constitutive relations may come in alternative forms as when Fick's law expresses the diffusive flux in terms of concentration gradients whereas the Maxwell relations for multicomponent diffusion express these gradients in terms of the fluxes. In many cases the forms can be converted into one another and one should keep an unprejudiced mind in case a conversion is desirable, but in some cases there are natural choices. Thus in a problem of diffusion and first-order reaction

$$D\nabla^2 c = kc \text{ in } \Omega$$

$$D\underset{\sim}{n} \cdot \nabla c = h(c_f - c) \text{ on } \partial\Omega$$

one could formulate the problem in terms of a flux

$$j = -D\nabla c$$

since then

$$c = -\frac{1}{k} \nabla \cdot j$$

and

$$D\nabla^2 \cdot j = kj \text{ in } \Omega$$

with

$$-n \cdot j = h + (h/k)\nabla \cdot j \text{ on } \partial\Omega.$$

Though the second set of equations has three components and is clearly not the set to use to solve the problem yet the complementary formulations have their roles in the variational properties of the solutions since the solution c minimizes the functional

$$\iiint_\Omega \left[D(\nabla c)^2 + kc^2 \right] dV + \iint_{\partial\Omega} h \left[(c_f - c)^2 \right]$$

whilst j maximizes

$$2 \iint_{\partial\Omega} (n \cdot j) dS - \iiint_\Omega \left[\frac{1}{D} j^2 + \frac{1}{k} (\nabla \cdot j)^2 \right] dV - \iint_{\partial\Omega} \left[\frac{1}{h}(n \cdot j)^2 \right] dS$$

Many examples of this duality are to be found in Arthurs [19].

Powerful general principles can often be brought to bear on constitutive relations to show what general form they must have. Thus Serrin [160] shows that the stress tensor, T, of a Stokesian fluid must be related to the rate of strain tensor, D, in the form

$$T = \alpha I + \beta D + \gamma D^2$$

where α, β and γ may be functions of the three invariants of D. Examples of this kind of reasoning abound in rational mechanics, as, for example, in Truesdell and Toupin's materpiece [179].

3.3 Discrete and continuous models.

At this point it is useful to discuss the alternative formulation of discrete and continuous models of physical problems. Each has its advantages and disadvantages but even here the distinctions are not absolute nor clean-cut, either in substance or in method, and may manifest themselves at one time as ordinary vs. partial differential equations, at another in linear algebraic vs. integral equations. Moreover when it comes to computation on a digital computer, the continuous necessarily becomes discrete. The terms 'lumped parameter' and 'distributed parameter' systems seem misguided for it is variables not parameters that are lumped (discrete) or distributed (continuous).

The word 'lumping', in spite of its ungainly overtones, is useful in describing the process by which a number of things are put together in one. This may result in replacing a continuous system by a discrete one. An example of this is the network thermodynamics of Oster, Perelson and Katchalsky [141], where problem of flow and transport in biological systems are treated by the ideas of electrical network theory. This converts parabolic equations into ordinary differential equations and elliptic into algebraic. An important discussion of lumping in the context of mono-molecular reactions has been made by Wei and Kuo [190]. This reduces a large system of species into a continuum of species and in this sense lumps them together. Lumping of this sort does replace a large number of equations by a single equation, but this is often an integro-differential equation rather than an ordinary differential equation. In fact the sense of the word has been turned around and what is being done is the distribution of a large number of discrete variables into a continuous variable--vigorously stirring out the lumps. These two processes are worth considering further

47

as they illustrate some of the subtleties of the relations between continuous and discrete models. Consider first the loss of heat of a wall to its environment--a thermal analogue of Oster's case of diffusion through a membrane. The figure below shows the physical picture with $T(x,t)$,

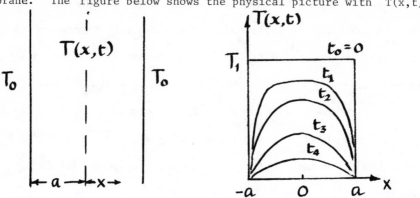

$-a \leq x \leq a$, the temperature at any point of the wall. Thus (incorporating the symmetry) the continuous model be

$$\rho c_p \frac{\partial T}{\partial t} = k \frac{\partial^2 T}{\partial x^2} , \ 0 < x < a,$$

$$\frac{\partial T}{\partial x} = 0, \ x = 0,$$

$$k \frac{\partial T}{\partial x} = h(T_o - T), \ x = a,$$

$$T(x,o) = T_1.$$

(3.6)

The kind of solution that we expect for these equations is sketched in the right part of this figure. Let us make the equations dimensionless by writing

$$\xi = x/a, \ \tau = kt/\rho c_p a^2, \ u = (T-T_o)/(T_1-T_o), \ \mu = ha/k.$$ (3.7)

Thus

$$u_\tau = u_{\xi\xi}, \quad 0 < \xi < 1,$$

$$u_\xi = 0, \qquad \xi = 0,$$

(3.8)

$$u_\xi + \mu u = 0, \xi = 1,$$

$$u(\xi, 0) = 1.$$

The complete solution of this equation is elementary enough, being

$$u(\xi, \tau) = \sum_1^\infty \frac{2\mu}{\lambda_n^2 (\sec\lambda_n + \sin\lambda_n)} e^{-\lambda_n^2 \tau} \cos\lambda_n \xi \qquad (3.9)$$

where

$$\lambda_n \tan\lambda_n = \mu, \quad n = 1, 2, \ldots \qquad (3.10)$$

In particular the average temperature is

$$\bar{u}(\tau) = \sum_1^\infty \frac{2\mu^2}{\lambda_n^2 (\lambda_n^2 + \mu\lambda_n + \mu^2)} e^{-\lambda_n^2 \tau}. \qquad (3.11)$$

The sequence of eigenvalues λ_1, λ_2 increases rapidly; for example with large μ, $\lambda_n = (2n-1)\pi/2$. Thus all the terms after the first quickly became negligible and

$$\frac{d}{d\tau} \bar{u}(\tau) \sim -\lambda_1^2 \bar{u}(\tau). \qquad (3.12)$$

This makes it look like a first order system with a time constant of λ_1^{-2}. In terms of real time this is

$$a^2/\lambda_1^2 D, \qquad (3.13)$$

or $4a^2/\pi^2 D$ as $\mu \to \infty$.

The network analogue consists in lumping the resistances and examining the driving potentials. A wall of unit area, thickness 2a and conductivity k will conduct heat at a rate $Q = k\Delta T/2a$ and might therefore be regarded as having a resistance of $\Delta T/Q = 2a/k$. Let this resistance which is in

49

fact distributed over the thickness, be divided into two and lumped at the surfaces of the slab as shown in the figure. The temperature within the slab sill also have to be lumped and since the resistance has been separated to the walls the natural lumping is $\bar{T}$, the mean temperature. The flow of

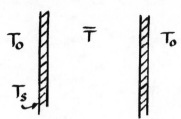

heat out of the slab at each face will therefore be $(k/a)(\bar{T}-T_s)$ where T_s is the temperature at the outside surface of the slab. This just matches the flow through the external film $h(T_s-T_o)$ so we have

$$q = \frac{k}{a}\,(\bar{T}-T_s) = h(T_s-T_o) = \frac{k}{a}\,(1+\tfrac{1}{\mu})^{-1}(\bar{T}-T_o).$$

Now the thermal capacity of the slab is clearly $2a\rho c_p$, so

$$2a\rho c_p \frac{d\bar{T}}{dt} = 2\,\frac{k}{a}\,(1+\tfrac{1}{\mu})^{-1}(T_o-\bar{T}) \tag{3.14}$$

or

$$\frac{d\bar{u}}{d\tau} = -(1+\tfrac{1}{\mu})^{-1}\,\bar{u} \tag{3.15}$$

This gives a dimensionless time constant of $(1+1/\mu)$ or in dimensional terms

$$a^2(1+\mu)/D\mu, \tag{3.16}$$

and in the limit $\mu \to \infty$,

$$a^2/D. \tag{3.17}$$

Comparing (3.13) and (3.17) we see that we are off by a factor of two and a half. This suggests that agreement could be improved by assuming that the "dynamic" capacity of the wall is only two fifths of its "static" capacity $2a\rho c_p$. For the more general case of external resistance the ratio of the time constants of the lumped and distributed systems is

$$t_L/t_D = \mu/(1 + \mu)\lambda_1^2$$

$$= \frac{\tan\lambda_1}{\lambda_1(1+\lambda_1\tan\lambda_1)}$$

As μ decreases from infinity to zero, λ_1 decreases from $\pi/2$ to 0 and the ratio increases from $(4/\pi^2)$ to 1. It is not surprising that the two time constants should agree in the limit $\mu \to 0$, or $k/a \ggg h$, for if the resistance of the outside films dominates so completely then the system is truly lumped.

This comparison illustrates some of the difficulties in going from the discrete to the continuous. A certain amount of accuracy can be recovered by the use of a pseudo-capacity, much as a virtual mass can be used in other cases, but it is not altogether satisfactory to have the wall capacity depend on the heat transfer coefficient. An analogous method of reducing the partial differential equations of a catalyst particle to ordinary differential equations was used by Hlavácek et al. [95]. Reference to this is given in [12] and [14] where the one-point collocation method of Villadsen and Stewart [188] is also described. (See also Sec. 4.5.2 below and [206]).

To illustrate the kind of "lumping" that is really distribution consider Luss and Hutchinson's [123] treatment of many parallel first order reactions. In many situations it is not possible to describe a mixture of chemical species that boil between, say, 350° and 500° and this might be taken as a largish lump. On the other hand if we talk about the number of moles $n(T)dT$ that boil in the range $(T, T + dt)$ we have really made a continuum, i.e. an infinity of species, out of a system that is necessarily discrete. In the case of species which can all undergo a reaction $A_i \to B_i$ with rate constant k_i we may devise a continuum and talk about the "species" $A(k)dk$ as all that reacts with rate constant in the range $(k, k + dk)$. If $c(t,k)dk$ is the concentration of this material at time t and if the reactions are all parallel first order

$$c(t,k) \ dk = c(o,k) \ dk \ e^{-kt} \qquad (3.18)$$

Now in many cases we may only be interested in the total amount $C(t) = \int_o^\infty c(t,k)dk$ and we see that

$$C(t) = \int_o^\infty e^{-kt} c(o,k)dk \qquad (3.19)$$

is the Laplace transform of the initial distribution with time, for a change playing the role of the transform variable.

It is interesting to enquire if there is an apparent rate law

$$C = \frac{dC}{dt} = -f(C) \qquad (3.20)$$

but although $\dot C = -\int_o^\infty k e^{-kt} c(o,k)dk$ it is seldom possible to invert (3.19) and so eliminate t. A notable exception is

$$c(o,k) = C(o)k^\alpha e^{-\beta k} \frac{\beta^{\alpha+1}}{\Gamma(\alpha+1)} \ , \ \alpha, \ \beta > 0. \qquad (3.21)$$

Let us make c dimensionless by dividing by $C(o)$, i.e.

$$u(t,k) = C(t,k)/\int_o^\infty c(o,k)dk \qquad (3.22)$$

so that for this distribution

$$u(o,k) = \beta^{\alpha+1}k^\alpha e^{-\beta k}/\Gamma(\alpha+1) \qquad (3.23)$$

Then by eqn. (3.19)

$$U(t) = \frac{C(t)}{C(o)} = (1 + \frac{t}{\beta})^{-(\alpha+1)} \qquad (3.24)$$

and

52

$$\dot{U}(t) = \frac{\alpha+1}{\beta}(1 + \frac{t}{\beta})^{-(\alpha+2)} = \frac{\alpha+1}{\beta} U^\gamma \qquad (3.25)$$

where

$$\gamma = (\alpha+2)/(\alpha+1)$$

Thus the "lump" appears to decay as a γth order reactant with γ depending on the parameter α in the initial distribution. It is not surprising that as $\alpha \to \infty$, $\gamma \to 1$ for the variance σ^2 of the initial distribution is $(\alpha+1)/\beta^2$ and the mean, $1/(\alpha+1)\mu$ is $(\alpha+1)\beta$. Thus $\alpha^2/\mu^2 = 1/(\alpha+1) \to 0$ as $\alpha \to \infty$, the distribution becomes narrower and therefore appears to decay in first order fashion. It is remarkable that for all α and β the rate constant in eqn. (3.25) is $\mu = (\alpha+1)/\beta$, the mean value of k in the initial distribution. If $\alpha = 0$ the apparent order is $\gamma = 2$ and it is noteworthy that second order reaction rates have been used to correlate hydrocarbon cracking for some time.

Though it is seldom possible to get complete results of this kind, Luss exploited the convexity of the exponential to show that

$$e^{-\mu t} < U(t) < \left[\sigma^2 + \mu^2 e^{-\nu t}\right]/\left[\sigma^2 + \mu^2\right],$$

$$\nu = (\sigma^2 + \mu^2)/\mu.$$

Such a result is extremely useful (and incidentally an excellent illustration of the value of the theory of inequalities) as it gives bounds on an observable in terms of certain calculable functionals, in this case the mean rate constant and their variance in the initial distribution. For an extended discussion of continuous mixtures see Gavalas and Aris [73], while for some alternative treatments Liu and Lapidus [118] and Bailey [21] may be consulted.

4 How should a model be manipulated into its most responsive form?

"You boil it in sawdust: you salt it in glue:

You condense it with locusts and tape:

Still keeping one principal object in view--

To preserve its symmetrical shape."

C. L. Dodson. The Hunting of the Snark.

Fit 5. St. 24.

4.1 Introductory suggestions.

Though a model may have been formulated with perfect propriety and perspicacity it is almost always a mistake to jump in with an extensive series of computations. It is better to live with it for a bit, to view it from different angles, to shape and mould it more justly. If the analogy may be permitted, there is a need for mathematical foreplay if model is to be fully responsive and the ultimate knowledge is to be satisfactory. The analogy is not inappropriate in that a like delicacy and tentativeness are required, but it breaks down in the bibliography for while the literature of the one art is enormous, though for the most part preternaturally dull, tha of the other, for all its excitement, is thin on the ground. Nor is the maxim of going to the masters as much help as it might be, for though, like Jacob's, their wrestling is before the breaking of the day, unlike him, the seldom show it in their gait. There is of course a considerable literature on intellectual creativity but that is not the issue here. Our needs are closer to those that have been met by Polya's examination of the art of problem solving [145-147] but in some ways are even preparatory to this.

Lin and Segel's excellent text is one of the very few that addresses itself to this question, though there is much to be learned from Hammersley's vivacious writing on the strategy and tactics of research in mathematics [80-82] and Noble's book [136].

Hammersley's "maxims for manipulators" [80] are worth summarizing since they give the flavour of his thinking and raise several points we shall return to later. They are: i, clean up the notation; ii, choose suitable units; iii, reduce the number of variables wherever possible; iv, draw rough sketches and examine particular cases; v, avoid rigour like the plague, it only leads to _rigor mortis_ at the manipulative stage; vi, have about an equal amount of stuff on each side of the equation. These he illustrates with a maximimization problem of deceptively innocent aspect. If this is tackled bullheadedly without the kind of prior manipulation Hammersley discusses it proves very resistant to accurate and apprehensible solution. Though he is obviously the poorer golfer brandishing a bigger bag of clubs, Aris' "maxims for mathematical modelling" [15] will perhaps bear repetition.

Maxims for Mathematical Modelling

1. Cast the problem in as elegant a form as possible.

2. Choose a sympathetic notation, but don't become too attached
    to it.

3. Make the variables dimensionless, since this is the only way
    in which their magnitudes take on general significance, but
    do not lose sight of the quantities which may have to be varied
    later on in the problem nor forget the physical origin of each
    part.

4. Use a priori bounds of physical or mathematical origin to keep all variables of the same order of magnitude, letting the dimensionless parameters show the relative size of the several terms.

5. Think geometrically. See when you can reduce the number of variables (even at the expense of first treating an over-simplified problem), but keep in mind the needs of the general case.

6. Use rough and ready methods, but don't carry them beyond their point of usefulness. (e.g. Isoclines in the phase plane.)

7. Find critical points and how the system behaves near them or what is asymptotic behaviour is at long or short times.

8. Check limiting cases and see how they tie in with simpler problems that can be solved explicitly.

9. Use crude approximations, e.g. 1-point collocation. Trade on the analogies they suggest, but remember their limitations.

10. Rearrange the problem. Don't get fixed ideas on what are the knowns and what the unknowns. Be prepared to work with implicit solutions.

11. Neglect small terms, but distinguish between regular and singular perturbations.

12. Use partial insights and despise them not. (e.g. Descartes rule of signs).

13. These maxims will self-destruct. Make your own!

Of these the most important is surely the last for they are not to be regarded as a book of rules with promise of success to any who will apply them. Rather they are suggestions and if anyone is to "think in the marrow bone" he must grow his own bone marrow.

The graith of the compleat mathematical modeller is a rich one and it is beyond the scope of these notes to survey the whole range of it, but we shall want to say something about the preliminaries of nomenclature and the choice of dimensionless variables. The possible reduction of the number of equations also deserves some attention, but the nub of the matter lies in the whole business of getting a feel for the solution without doing the detailed calculations. To see this done skillfully as, for example, by Paul Fife in his treatment of pattern formation [65], is to share the delight of the true craftsman. There are many tools that are available for this task and we shall only be able to touch on a few of them in any detail. However a preliminary list (in no particular order) would certainly include:

1. Obtaining a priori bounds on the solution;

2. Use of fixed point theorems and contraction maps;

3. Location of roots (Descartes' rule of signs);

4. Use of special properties (e.g. the constant cross-ratio of
    four solutions of Ricatti's equation);

5. Method of isoclines;

6. Degree theory;

7. Use of perturbation theory;

8. Asymptotic analysis, singular perturbation theory;

9. Group theory;

10. Use of inequalities;

11. Finding moments of the solution;

12. Intermediate asymptotics, travelling waves;

13. Integration by parts (e.g. Frisch's treatment of diffusion lag);

14. Maximum principles;

15. Numerical approximations (even crude ones like one-point collocation);

16. Variational principles;

17. Linearization;

18. Integral methods (e.g. in boundary layer theory).

4.2 Natural languages and notations.

Certain branches of mathematics lend themselves naturally to expressing the structure of a physical theory. It is well-known that the theory of relativity found a natural language in the calculus of tensors, or "Ricci-calcul" as it was first called. Similarly the language of Cartesian vectors and tensors is natural for mechanics and linear algebra for the stoichiometr of chemical reactions. Thus the physical realities are clearly represented by the vectorial equation $m\ddot{\underset{\sim}{r}} = f$ and though it may be necessary to use the equations in component form, $m\ddot{x} = f_x$, etc., one should never lose sight of the whole in dealing with the parts. The notational convention of having one symbol for the vector or tensor helps in this. When a component notatio is used such rules as the summation convention are also valuable and the compactness of $\underset{\sim}{1} . \underset{\sim}{x}$ or $1_i x^i$ has advantage over $1_1 x_1 + 1_2 x_2 + 1_3 x_3$ or $1x + my + nz$.

Hammersley's first maxim (to clean up the notation) has more to it than meets the eye, for the right choice of notation, though in some respects a matter of personal taste, is not to be dismissed as trivial. It is easy enough to make mistakes in manipulation under the best of circumstances but to burden oneself with a cumbersome and unsuitable notation is downright stupid. Except where some overriding convention has a prerogative, the system for the basic physical quantities should be generally as simple and mnemonic as possible. It is not always possible to avoid suffixes but again

58

it is important that they should be natural. Thus in the example of App. C

by calling the species $A_1, A_2 \ldots A_S$, it is natural to denote their concen-

trations by c_j, j = 1,2...S. Similarly at the early stage it is better to

call the volume of the coolant V_c than to preempt yet another letter. How

sympathetic one person's notation may appear to another is a matter of

context and personal taste. For example chemical engineering texts often

use Θ for the residence or holding time of the stirred tank, V/q, while

biochemical engineers use D for its reciprocal, calling it the dilution

rate; yet, since D plays the role of a death rate, many would feel that

Θ itself would be preferable. In choosing symbols for the dimensionless

variables it is seldom possible to make a consistent translation between

Latin and Greek letters, though it is an admirable tradition to follow.

Thus coordinates (x,y,z) can often become (ξ, η, ζ) or time t become

dimensionless time τ. Both alphabets tend to run out all too quickly, but

it is better to have recourse to Ps. 119 in the King James than to succumb

to such symbolic solecisms as V for a diffusion coefficient or x for a

mass. In general single letters should be used for a single quantity, but

tradition has sanctified the union of a capital and lower case for the

notable dimensionless groups such as the Reynolds number, Re. This seldom

causes confusion and is less unsightly than the convention N_{Re}. Chemical

engineers seem to have led the field in coining names for dimensionless

groups and hence immortalizing one another--much as the naturalists of the

18th centruy did with the taxonomy of Linnaeus--though it may be questioned

whether it helps to perpetuate Damköhler's memory in four numbers with Roman

numeral suffixes. Referring again to the example of App. C, the concentra-

tions and temperatures, c_j, T, T_w and T_c become u_j, v, w and Θ in

their dimensionless forms. The last choice is not altogether a happy one, but v_c has to be avoided if we are to have ultimate freedom from suffixed for the dependent variables. In the case of Σ_6 whose reduction to dimensionless form is to be studied in detail in the next section we shall use v_c for the dimensionless coolant feed temperature. This however is a fixed constant, not a dependent variable, and we need Θ as the conventiona symbol for residence time.

4.3 Rendering the variables and parameters dimensionless.

If the matter of notation is incidental, being governed by the canons of taste and common sense rather than high principle, the matter of making the variables dimensionless is of the essence. A physical magnitude has meaning with respect to an arbitrary set of standards and two quantities of the same dimensions measured in the same units can be compared. But even the world-wide adoption of SI units cannot give it any intrinsic meaning and the current effort in metricization, while useful to trade and engineering, has considerably less to contribute to science than would accrue from the revival of Latin as a lingua franca. It is only when quantities are made dimensionless that their magnitudes acquire an intrinsic meaning in the context of the model.

It might be worth remarking parenthetically that we are here practising so-called "inspectional analysis" [259,260], that is, the manipulation of the basic equations to reveal the dimensionless groups. It is the subtlest and most penetrating way of the methods sometimes called dimensional. Dimensional analysis, as expounded by Bridgeman [203] and others, relies on Rayleigh's method of indices [236] and the Buckingham Pi theorem to excogitate the dimensionless groups from a list of the significant physical quantities. The Pythagorean method, to use the name for the method of ratios coined by Becker in his excellent little monograph [201], refers to the technique of forming dimensionless ratios from the significant physical quantities peculiar to the system. Becker distinguishes three subgroups of dimensionless ratios: ratios of fluxes (e.g. of momentum energy or matter), which are appropriately called Heraclitian; geometrical ratios (e.g. aspect ratio), called the Pythagorean subgroup; and ratios of amounts or charges, called Democritian in recognition of the fact that the conceptual separation of matter from space goes back to Democritus and the atomists. He uses the term "configuration analysis" for the incorporation of the Pythagorean method into a larger analysis of physical systems and claims that it is "immensely more persuasive than dimensional analysis, almost as penetrating as inspectional analysis, and more versatile than either of these." The whole subject of dimensional analysis has a considerable literature; cf. [202,203,207-9,212,215-8,220-1,229-33,236-9,247,259,260].

The principle of making the equations dimensionless is simple enough. Each variable, dependent or independent, is expressed as a product of some characteristic quantity of the same dimensions and a dimensionless variable. The equations are then rearranged until a suitable set of dimensionless parameters appears. Two principles govern this process: i, constant quantities should be used as characteristic quantities; ii, the dimensionless parameters should bear the burden of showing the comparative importance of the various terms in the equation.

Let us try to elucidate this by considering the model Σ_6 of App. C the stirred tank with a single first-order irreversible reaction. There are usually several different ways of rendering equations non-dimensional and of choosing the dimensionless parameters and their pros and cons are best illustrated by an example. If we put $k(T) = A\exp(-E/RT)$ in eqns. (C15) and (C16) and replace hA in (C16) by h (to avoid confusion between the area and the pre-exponential factor in k), the equations of this model are:

$$V \frac{dc}{dt} = q(c_f - c) - VAe^{-E/RT}c \tag{4.1}$$

$$VC_p \frac{dT}{dt} = qC_p(T_f - T) - h(T - T_{cf}) + (-\Delta H)VAe^{-E/RT}c \tag{4.2}$$

Note that the same letters will be used for the same kind of parameters in the different forms of the equation but they will be differently related to the physical constants in the different modes of non-dimensionalizing. They are always defined in context.

   Let c*, T* and t* be a characteristic concentration, temperature and time to be chosen later and

$$\tau = t/t^*, \quad u = c/c^*, \quad v = T/T^* \tag{4.3}$$

and, by the persistence and simplification of indices,

$$u_f = c_f/c^*, \quad v_f = T_f/T^*, \quad v_c = T_{cf}/T^*. \tag{4.4}$$

Multiplying the first equation by t*/Vc* gives

$$\frac{du}{d\tau} = \frac{qt^*}{V}(u_f - u) - At^* u e^{-E/RT^* v} \tag{4.5}$$

There are three parameters qt*/V, At* and E/RT*. Now V/q is the so-called residence time so that V/qt* is the dimensionless residence time, say Θ, At* is a measure of the reaction rate of Damköhler number, say α, and E/RT* is the dimensionless activation energy or Arrhenius number, say γ. Thus

$$\frac{du}{d\tau} = \frac{u_f - u}{\Theta} - \alpha u e^{-\gamma/v} \tag{4.6}$$

   The second equation must be multiplied by t*/VC_pT* to give

$$\frac{dv}{d\tau} = \left[\frac{qt^*}{V}v_f + \frac{ht^*}{VC_p}v_c\right] - \left[\frac{qt^*}{V} + \frac{ht^*}{VC_p}\right]v + \frac{(-\Delta H)c^*}{C_p T^*}At^* u e^{-\gamma/v}$$

62

From this it is clear that $(-\Delta H)c*/C_p T* = \beta$, say, is an important parameter; moreover it has the nature of a dimensionless temperature rise due to reaction for $(-\Delta H)c*$ is the heat released by the complete reaction of the feed at the characteristic concentration. For the moment let us not reduce everything completely but set

$$\beta = \frac{(\Delta H)c*}{C_p T*}, \quad \delta_1 = \frac{qt*}{V} = \frac{1}{\theta}; \quad \delta_2 = \frac{ht*}{VC_p}, \quad \delta = \delta_1 + \delta_2 \tag{4.7}$$

Then

$$\frac{dv}{d\tau} = (\delta_1 v_f + \delta_2 v_c) - \delta v + \alpha\beta u e^{-\gamma/v} \tag{4.8}$$

$$= \delta(\bar{v}-v) + \alpha\beta u e^{-\gamma/v}$$

if we define

$$\bar{v} = (\delta_1 v_f + \delta_2 v_c)/\delta \tag{4.9}$$

as the weighted mean of the feed and coolant temperatures.

We now have various directions in which we can go. First suppose that all conditions are to be held constant and that we are interested in reducing the number of parameters to the minimum. Then in the equation for u we can make u_f, γ and θ all equal to 1 by choosing

$$c* = c_f, T* = E/R, \quad t* = V/q \tag{4.10}$$

giving

$$\frac{du}{d\tau} = 1 - u - \alpha u e^{-1/v} \tag{4.11}$$

No amount of manipulation will get rid of α. In the equation for v there appear to be five additional constants δ_1, δ_2, v_f, v_c and β, but $\delta_1 = 1/\theta = 1$ since all conditions are constant v_c and v_f can be condensed into $\bar{v}$ by eqn. (4.9). It is well to take one constant over to the left since this will be eliminated when we set the derivatives equal to zero to study the steady state. Thus let $\bar{v}$ be defined as above and

$$\zeta = \beta/\delta \tag{4.12}$$

so that

$$\frac{1}{\beta}\frac{dv}{d\tau} = \frac{\bar{v}-v}{\zeta} + \alpha u e^{-1/v} \tag{4.13}$$

Between equations (4.11) and (4.13) there are four parameters

$$\alpha = Av/q, \quad \beta = (-\Delta H)Rc_f/C_p E, \quad \zeta = [((-\Delta H)Rc_f/EC_p]/[1+h/qC_p] \tag{4.14}$$

and

$$\bar{v} = \frac{R(qC_p T_f + hT_{cf})}{E(qC_p + h)} \tag{4.15}$$

The steady state equations however contain only three, α, ζ and $\bar{v}$. They can be combined into a single equation by substituting for u from eqn. (4.11) in eqn. (4.13),

$$v-\bar{v} = \zeta - \frac{\alpha e^{-1/v}}{1+\alpha e^{-1/v}} \tag{4.16}$$

This equation is studied later in Sec. 4.5. The points to be made in favour of this mode of non-dimensionalizing are that it does give us the simplest possible form of the equations, for example, there is no parameter in the exponential function. The objection that can be raised to it is that the magnitudes of u and v are generally very different; u is between 0 and 1 but if E/R is of the order of $10000°$, as it may well be, even a fairly high temperature such as $500°K$ makes v of the order of 0.05.

A second method--and one that overcomes this difficulty--is to let u_f, v_f and Θ all be equal to 1, by putting

$$c^* = c_f, \quad T^* = T_f, \quad t^* = V/q. \tag{4.17}$$

Then

64

$$\frac{du}{d\tau} = 1 - u - \alpha u e^{-\gamma/v} \tag{4.18}$$

$$\frac{dv}{d\tau} = (\delta_1 + \delta_2 v_c) - \delta v + \alpha \beta u e^{-\gamma/v} \tag{4.19}$$

We thus have

$$\alpha = AV/q, \quad \beta = (-\Delta H)c_f/C_p T_f, \quad \gamma = E/RT_f, \quad \delta_1 = 1 \tag{4.20}$$

and

$$\delta_2 = h/qC_p$$

If T_f and T_c are not too disparate it is advantageous to put

$$T^* = (qC_p T_f + hT_{cf})/(qC_p + h) \tag{4.21}$$

since then the second equation becomes

$$\frac{1}{\delta}\frac{dv}{d\tau} = 1 - v + \zeta \alpha u e^{-\gamma/v} \tag{4.22}$$

where

$$(1/\delta) = qC_p/(qC_p + h), \quad \zeta = qc_f(-\Delta H)/(qC_p T_f + hT_{cf}) \tag{4.22}$$

A variant of the second method considers u and v to be the deviations of c and T from c_f and T_f by setting

$$u = (c_f - c)/c_f, \quad v = (T - T_f)/T^*. \tag{4.23}$$

Before choosing T^* let us see what this does to the exponential function. Since $T = T_f + T^* v$, $\exp-(E/RT) = \{\exp-(E/RT_f)\}\{\exp E(T-T_f)/RTT_f\} = \exp-(E/RT_f)\exp\{ET^*v/RT_f^2(1+T^*v/T_f)\}$. Thus if we choose $T^* = RT_f^2/E$ and put $\gamma = E/RT_f$ we have

$$k(T) = Ae^{-E/RT} = k(T_f)\exp[v/(1+v/\gamma)]. \tag{4.24}$$

The choice has two points in its favour. In the first place, it brings $k(T_f)$ out as a factor and this with c_f gives $k(T_f)c_f$ which is the reaction rate at feed conditions. In fact the equations are now:

$$\frac{du}{d\tau} = -u + \alpha(1-u)\exp[v/(1+v/\gamma)],\qquad(4.25)$$

$$\frac{1}{\delta}\frac{dv}{d\tau} = -v + v_c + \alpha\zeta(1-u)\exp[v/(1+v/\gamma)],\qquad(4.26)$$

where $\alpha = Vk(T_f)/q$, $\beta = (-\Delta H)c_f E/C_p RT_f^2$, $\gamma = E/RT_f$, $\delta = (qC_p + h)/qC_p$, $\zeta = \beta/\delta$ and $V_c = hE(T_{cf}-T_f)/RT_f^2(h + qC_p)$. Again, if the weighted mean feed temperature $(qC_p T_f + hT_{cf})/(qC_p + h)$ were used everywhere in place of T_f, we would have the same equations but $V_c = 0$. The second advantage of this form is that it makes natural the approximation of the exponential function by e^v, an approximation which can be obtained formally by letting $\gamma \to \infty$.

Finally—for it would be possible to play a large number of variations on this simple theme—consider the situation when an important physical quantity that affects more than one parameter, for example, q, is to be varied. If neither limiting case, $q \to 0$ or $q \to \infty$, is to be considered there is no real objection to retaining $t^* = V/q$ since the equations are autonomous. If however we wish to avoid this we can take $t^* = VC_p/h$ giving $\theta = h/qC_p$, $\alpha = AVC_p/h$, $\delta_1 = 1/\theta$, $\delta_2 = 1$. The reference temperature should not be taken to be the weighted mean as in eqn. (4.21) since this will vary with q but we can take $v = E(T-T_f)/RT_f^2$. Thus, for example, with

$$u = (c_f-c)/c_f,\quad v = E(T-T_f)/RT_f^2,\quad v_c = E(T_{cf}-T_f)/RT_f^2,\qquad(4.27)$$

$$\alpha = AVC_p/h,\quad \beta = (-\Delta H)Ec_f/C_p RT_f^2\qquad(4.28)$$

we have

$$\theta\frac{du}{d\tau} = -u + \alpha\theta(1-u)\exp[v/(1+v/\gamma)]\qquad(4.29)$$

$$\theta\frac{dv}{d\tau} = \theta v_c - (1+\theta)v + \alpha\beta\theta(1-u)\exp[v/(1+v/\gamma)].\qquad(4.30)$$

66

In this form the equations have parameters α, β, γ, v_c and Θ, the last being the variable parameter. This is the case extensively studied by Uppal, Ray and Poore [184], though they took a slightly different dimensionless time the effect of which was to confine the region of multiplicity to $0 < \Theta < 1$.

Another question that may be raised about any particular choice of dimensionless variables is whether it fits in with a larger scheme. For example if we replace T_{cf} in eqn. (4.2) by T_c and append a third equation of the form of (C 14)

$$v_c C_{pc} \frac{dT_c}{dt} = q_c C_{pc}(T_{cf}-T_c) + h(T-T_c) \tag{4.31}$$

we have three equations for c, T and T_c. Consider the extension of the form of eqns. (4.25) and (4.26) by setting

$$u = (c_f-c)/c_f, \quad v = (T-T_f)E/RT_f^2, \quad w = (T_c-T_f)E/RT_f^2 \tag{4.32}$$

with

$$t^* = V/q, \quad \alpha = Vk(T_f)/q, \quad \beta = E(-\Delta H)c_f/RT_f^2 C_p, \quad \gamma = E/RT_f,$$

$$\delta_1 = 1, \quad \delta_2 = h/qC_p, \quad \omega = V_c C_{pc}/VC_p, \quad \chi = q_c C_{pc}/qC_p. \tag{4.33}$$

Then we have

$$\frac{du}{d\tau} = -u + \alpha(1-u)\exp[v/(1+v/\gamma)], \tag{4.34}$$

$$\frac{1}{\delta}\frac{dv}{d\tau} = -v + \frac{\delta_2}{\delta} w + \alpha\beta(1-u)\exp[(1+v/\gamma)], \tag{4.35}$$

$$\omega \frac{dw}{d\tau} = \chi(\dot{w}_f-w) + \delta_2(v-w), \tag{4.36}$$

a set of three equations with two additional parameters ω and χ the static and fluent heat capacity ratios.

A similar example which we will take up from other aspects later can be given rather briefly at this point. It concerns the growth of two organisms in a chemostat--the microbiologist's name for a stirred tank reactor. Two organisms, species 1 and 2, are present in concentrations, X_1 and X_2 and feed on a common nutrient whose concentration is S in a vessel whose volume is $(1/D)$ times the flow rate, i.e. $q = DV$. (This is the standard notation of biochemical engineering and D is known as the dilution rate). They grow at rates μ_1 and μ_2 which are functions of S and one moiety of S yields Y_1 and Y_2 moieties of the two species respectively. Thus

$$\frac{dX_1}{dt} = \mu_1(S)X_1 - DX_1 \qquad (4.37)$$

$$\frac{dX_2}{dt} = \mu_2(S)X_2 - DX_2 \qquad (4.38)$$

$$\frac{dS}{dt} = \frac{\mu_1(S)X_1}{Y_1} - \frac{\mu_2(S)X_2}{Y_2} + D(S_F - S) \qquad (4.39)$$

where only nutrient is fed to the chemostat and at a concentration S_F. The functions $\mu_i(S)$ are of the form

$$\mu_i(S) = M_i \left\{ 1 + \frac{K_i}{S} + \frac{S}{L_i} \right\}^{-1} \qquad (4.40)$$

We want to investigate the way in which the system depends on S_F and D so it will not do to use these values to make the concentrations and time dimensionless. Without loss of generality we can give the label 1 to the species which has the greater growth rate for large S, i.e. $M_1 L_1 \geq M_2 L_2$. Then $1/M_1$ has the dimensions of time and K_1 of concentration and accordingly we let

$$x = X_1/Y_1 K_1, \quad y = X_2/Y_2 K_1, \quad z = S/K_1, \quad \tau = M_1 t. \qquad (4.41)$$

There are then four fixed parameters

$$\alpha = M_2/M_1, \quad \beta = K_2/K_1, \quad \gamma = L_2/K_1, \quad \delta = L_1/K_1 \qquad (4.42)$$

and two we wish to vary

$$\Theta = D/M_1, \quad Z = S_F/K_1. \qquad (4.43)$$

Then

$$x = \{g_1(z) - \Theta\}x$$

$$y = \{g_2(z) - \Theta\}y \qquad (4.44)$$

$$z = \Theta(Z-z) - xg_1(z) - yg_2(z)$$

where

$$g_1 = \left\{1 + \frac{1}{z} + \frac{z}{\delta}\right\}^{-1}, \quad g_2 = \alpha\left\{1 + \frac{\beta}{z} + \frac{z}{\gamma}\right\}^{-1} \qquad (4.45)$$

We remark that γ and δ can be infinite if there is no substrate inhibition.

4.4 Reducing the number of equations and simplifying them.

The example we have been considering extensively in the last section was a model, Σ_6, with just two equations. It had gotten to be that way by first reducing the model to one described by $S+1$ equations, (one for the concentration of each species and one for the temperature) and then observing that, since the reaction rate depended on only one of the concentrations, only one equation was needed in the concentration and one in the temperature. However for constant feed conditions a similar reduction can be made for the completely general reaction $\Sigma\alpha_j A_j = 0$, with reaction rate $r(c_1, c_2 \ldots c_S, T)$. For substitute

$$c_j = (q_f c_{jf}/q) + \alpha_j \xi \qquad (4.46)$$

in the equation

$$V \frac{dc_j}{dt} = q_j c_{jf} - qc_j + \alpha_j Vr(c_1, \ldots c_S, T) \tag{4.47}$$

and α_j comes out as a factor leaving

$$V \frac{d\xi}{dt} = -q\xi + Vr(\xi, T) \tag{4.48}$$

where

$$r(\xi, T) = r(\xi, T; q_1, \ldots q_r, c_{1f} \ldots c_{Sf})$$
$$= r((q_1 c_{1f}/q) + \alpha_1 \xi, \ldots, T) \tag{4.49}$$

Thus a single equation has been obtained for a single reaction. But it will be objected that the initial values $c_j(0) = c_{jo}$ may not all be expressible in the form

$$c_{jo} = (q_j c_{jf}/q) + \alpha_j \xi_o. \tag{4.50}$$

If they can all be so expressed then there is no problem for ξ_o then becomes the initial value for eqn. (4.48), but if they cannot then at least two values of

$$\eta_j = \{c_{jo} - (q_j c_{jf}/q)\}/\alpha_j \tag{4.51}$$

are different. Then let

$$c_j = (q_j c_{jf}/q) + \alpha_j(\xi + \eta_j \zeta) \tag{4.52}$$

a form which can take care of the initial conditions by letting $\xi(0) = 0$, $\zeta(0) = 1$. Substituting eqn. (4.52 in eqn. (4.38) and dividing through by α_j gives

$$\left[V \frac{d\xi}{dt} + q\xi - Vr(\xi, \zeta, T) \right] + \eta_j \left[V \frac{d\zeta}{dt} + q\zeta = 0 \right] \tag{4.53}$$

Now each of the expressions in brackets must be separately zero for otherwise all the η_j would be the same. The first bracket gives eqn. (4.48) except that now the reaction rate is a function of ζ as well as of ξ and T

70

and, parametrically, of c_{jo} as well as of c_{jf} and q_j. However the equation

$$V \frac{d\zeta}{dt} = -q\zeta \qquad (4.54)$$

has the immediate solution $\zeta(t) = \exp - qt/V$ and quickly tends to zero. We say that the feed and initial compositions are compatible if one can be derived from the other by some degree of reaction, i.e. they are related by eqn. (4.50). The ζ is a measure of the incompatibility of the current composition and the feed composition and it is very satisfying to see that this just "washes out" of the reactor by a purely physical equation without any reactive term as the memory of an initial composition should.

In a more general situation there might be R simultaneous reactions

$$\sum_{j=1}^{S} \alpha_{ij} A_j = 0, \quad i = 1, 2, \ldots R, \qquad (4.55)$$

which without loss of generality we can take to be independent, i.e. the $R \times S$ matrix with entries α_{ij} is of rank R. Then the mass balance over the jth species gives

$$V \frac{dc_j}{dt} = q(\bar{c}_{jf} - c_j) + V \sum_{i=1}^{R} \alpha_{ij} r_i (c_1, \ldots c_S, T) \qquad (4.56)$$

where $q\bar{c}_{jf} = q_j c_{jf}$. The substitution

$$c_j = \bar{c}_{jf} + \sum_{i=1}^{R} \alpha_{ij} \xi_i + (c_{jo} - \bar{c}_{jf})\zeta \qquad (4.57)$$

gives

$$\sum_{i=1}^{R} \alpha_{ij} \left[V \frac{d\xi_i}{dt} + q\xi_i - V r_i (\xi_1, \ldots \xi_R, \zeta T) \right]$$

$$+ (c_{jo} - \bar{c}_{jf}) \left[V \frac{d\zeta}{dt} + q\zeta \right] = 0.$$

The independence of the reactions and the incompatibility of the feed and initial compositions force each of the bracketed expressions to be severally zero and so give R+1 equations (or R, if the feed and initial compositions are compatible) in place of S.

When the reactor is adiabatic, a combination of concentration and temperature obeys a similarly simple equation. For, taking $h=0$ in eqn. (4.20) gives $\delta = \delta_1 = 1$, $\delta_2 = 0$, so that multiplying eqn. (4.18) by and adding eqn. (4.19) gives

$$\frac{d}{d\tau} (\beta u + v) = (1+\beta) - (\beta u + v). \tag{4.58}$$

This notion can be carried over to distributed systems as when the reaction $\Sigma \alpha_j A_j = 0$ takes place at a rate r per unit volume in a porous pellet, Ω, through which the reactants diffuse with effective Knudsen diffusion coefficients D_j. For this system we have the equations

$$D_j \nabla^2 c_j + \alpha_j r(c_1, \ldots, T) = 0 \text{ in } \Omega \tag{4.59}$$

with

$$c_j = c_{js} \text{ on } \partial\Omega. \tag{4.60}$$

This implies that each linear combination $(D_j c_j / \alpha_j) - (D_k c_k / \alpha_K)$ satisfies

$$\nabla^2 \{ (D_j c_j / \alpha_j) - (D_k c_k / \alpha_k) \} = 0 \text{ in } \Omega \tag{4.61}$$

and is constant over $\partial\Omega$. But a potential function constant on $\partial\Omega$ is constant everywhere in Ω and hence we can substitute

$$c_j = c_{js} + (\alpha_j / D_j)\xi \tag{4.62}$$

to give

$$\nabla^2 \xi + r(c_{1s} + (\alpha_1 / D_1)\xi, \ldots, T) = 0 \text{ in } \Omega, \tag{4.63}$$

with

$\xi = 0$ on $\partial \Omega$.

The energy balance gives

$$k_e \nabla^2 T = (\Delta H) r(c_1, \ldots T) \text{ in } \Omega \tag{4.64}$$

with

$$T = T_s \text{ on } \partial \Omega$$

and this suggests the substitution

$$T = T_s + (-\Delta H/k_e)\xi. \tag{4.65}$$

Thus all the equation collapse into one, namely

$$\nabla^2 \xi = -R(\xi) = -r(c_{1s} + (\alpha_1/D_1)\xi, \ldots, T_s + (-\Delta H/k_e)\xi) \text{ in } \Omega \tag{4.66}$$

with

$$\xi = 0 \text{ on } \partial \Omega.$$

Notice, however, that this is of restricted application. Except for symmetric regions, such a sphere, it does not apply with the more general boundary conditions

$$D_j \frac{\partial c_j}{\partial n} = k_j (c_{jf} - c_j) \text{ on } \partial \Omega.$$

Nor can it be extended to transients except in the special case of $(D_j \rho c_p/k_e) = 1$ for all j. In this case we write the transient equations as

$$\frac{\partial c_j}{\partial t} = D_j \nabla^2 c_j + \alpha_j r \; ; \quad \rho c_p \frac{\partial T}{\partial t} = k_e \nabla^2 T + (-\Delta H) r \tag{4.67}$$

and assume the initial values are uniform and c_{jo}, T_o respectively. Then putting

$$c_j = c_{js} + (\alpha_j/D_j)\xi + (c_{jo} - c_{js})\zeta$$

$$T = T_s + (-\Delta H/k_e)\xi + (T_o - T_s)\zeta \tag{4.68}$$

and letting

73

$$D = D_j = k_e/\rho c_p$$

gives

$$\left[\frac{\partial \xi}{\partial t} - D\nabla^2 \xi - R\right] + \frac{(c_{jo} - c_{js})D_j}{\alpha_j}\left[\frac{\partial \zeta}{\partial t} - D\nabla^2 \zeta\right] = 0. \tag{4.69}$$

Also $\xi = \zeta = 0$ on $\partial \Omega$ and $\xi = 0$, $\zeta = 1$ initially. If all the quantities $(c_{jo} - c_{js})D_j/\alpha_j$ are the same then we can fix ξ_o and ignore ζ. But if they are different then the same argument as before shows that the two brackets are severally equal to zero. Thus ζ is the distribution of temperature in Ω which is initially uniform and has zero boundary values and such a temperature, as we know, subsides to zero everywhere.

When such a reduction can be made the stability picture is gained from the reactive equation and the intrinsically stable equation for ζ adds nothing to the analysis of stability. For example, if

$$\frac{du}{d\tau} = 1 - u - \alpha R(u,v)$$

$$\frac{dv}{d\tau} = 1 - v + \alpha \beta R(u,v) \tag{4.70}$$

and $w = \beta u + v$, then an equivalent system is

$$\frac{du}{d\tau} = 1 - u - \alpha R(u, w - \beta u)$$

$$\frac{dw}{d\tau} = 1 + \beta - w \tag{4.71}$$

The linearization of the first about a steady state (u_s, v_s) gives

$$\begin{bmatrix} \dot{x} \\ \dot{y} \end{bmatrix} = \begin{bmatrix} -(1 + \alpha R_u) & -\alpha R_v \\ \alpha \beta R_u & -(1 - \alpha \beta R_v) \end{bmatrix} \begin{bmatrix} x \\ y \end{bmatrix} \tag{4.72}$$

where $x = u - u_s$, $y = v - v_s$. Similarly, if $z = w - w_s$, the linearization of eqns. (4.71) gives

74

$$\begin{bmatrix} \dot{x} \\ \dot{z} \end{bmatrix} = \begin{bmatrix} -(1 + \alpha R_u - \alpha\beta R_v) & -\alpha R_v \\ \cdot & -1 \end{bmatrix} \begin{bmatrix} x \\ z \end{bmatrix} \tag{4.73}$$

The stability criteria for these two matrices are the same.

The example of competing organisms that we intend to take up again later can be done in the short order now. The addition of the three equations in (4.44) gives

$$\dot{x} + \dot{y} + \dot{z} = \Theta\{Z - (x+y+z)\} \tag{4.74}$$

Thus if the point (x,y,z) does not lie in the plane

$$x + y + z = Z \tag{4.75}$$

it rapidly approaches it since

$$x + y + z = Z + \{x_o + y_o + z_o - Z\}e^{-\Theta\tau}$$

From this we see that all the steady states must lie in the plane, ABC of the figure below. If M is a starting point in ABC the trajectory of

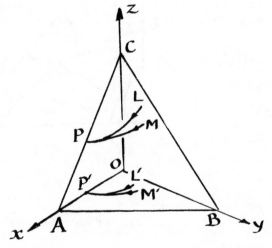

the solution of eqn. (4.44) lies wholly in the plane and approaches the steady state. The trajectory from a nearby point L, not in the plane, quickly approaches the trajectory MP and becomes tangential to it, if, as is the case shown here, the other eigenvalues are negative and greater than

Θ. This relationship is preserved in the projections L'P', M'P' on the plane OAB. It follows that we shall get the essence of the behaviour of the system by restricting attention to the plane

$$z = Z - x - y.$$

In this case we have the equations

$$\dot{x} = \{g_1(Z - x - y) - \Theta\}x$$

$$\dot{y} = \{g_2(Z - x - y) - \Theta\}y$$

For a certain combination of $\alpha, \beta, \gamma, \delta, \Theta$ and Z the phase plane of solutions of these two equations looks like this:

There are two stable steady states O and Q. Any initial state in the triangular region OPR is attracted to O, while all trajectories starting outside, i.e. in PRBA, go to Q. Thus RP is a separatrix between the two regions of attraction. It is the projection of the curve RP in the plane ABC, shown in the

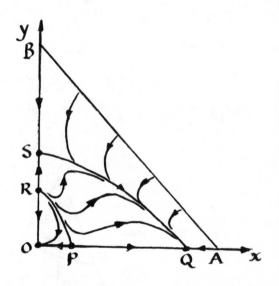

next figure (N.B. it is not P"R") onto OAB. The surface P'R'RR"P"P is the separatrix surface in space and allows for an arbitrary starting point. Thus any trajectory starting between this surface and the z-axis goes to O; all other trajectories go to Q. For some purposes we may need the whole separatrix surface but its structure is often sufficiently clear from its intersection with the plane.

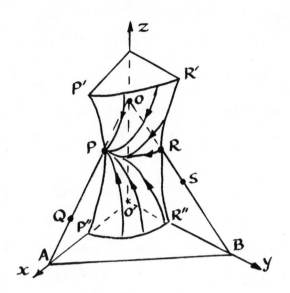

4.5 Getting partial insights into the form of the solution.

In the introductory paragraph to this chapter a number of techniques for getting some feel for the form of the solution were listed. This is not the place to explore them all in detail, but it is important to see some of them in action since they are at the heart of the whole craft of modelling. We shall do this by considering three examples. The first is an illustration of phase plane techniques in the chemostat that we have just been looking at; the second is the use of a coarse numerical method to suggest an approach to the uniqueness problem; the third is intended to show how far one can get toward a complete picture without any calculation.

4.5.1 The phase plane and competing populations.

The pair of equations

$$\dot{x} = \{g_1(Z - x - y) - \Theta\}x \qquad (4.76)$$

$$\dot{y} = \{g_2(Z - x - y) - \Theta\}y \qquad (4.77)$$

can have four types of steady state which can be obtained by setting $\dot{x} = \dot{y} = 0$. They bear the following interpretations and we shall use the notation shown at the left.

0	$x_s = y_s = 0$	complete wash-out of both x_1 and x_2
P or P and Q	$g_1(Z-x_s) = \Theta, \ y_s = 0$	x_1 grows, x_2 washes out
R or R and S	$x_s = 0, \ g_2(Z-y_s) = \Theta$	x_1 washes out, x_2 grows
T or T and U	$x_s + y_s = \zeta, g_1(Z-\zeta) = g_2(Z-\zeta) = \Theta$	both x_1 and x_2 coexist but in indeterminate proportions

The last possibility only arises if the curves $g_1(z)$ and $g_2(z)$ cross one another and then Θ must be chosen to have their common value. If the curves g_1 and g_2 are disposed as below and the point (Z,Θ) is at 0, then there are five steady states named according to the rule that proceeding to the left from 0 the first intersection with g_1 is P and the second, if there is one, is called Q; likewise the first and second intersections with g_2 are respectively R and S. Reading from right to left this disposition is uniquely described by the 'word' OPRSQ. In the x,y-plane this is also the order of increasing distance from the origin though P and Q are of course on the x-axis whilst R and S are on the y-axis. We can put the steady states in part (b) of the preceding figure. The 45° diagonals have been drawn in not only to confirm the order of placing the points but also with the reminder that they are isoclines of horizontal and vertical passage. Thus $\dot{x}=0$ on the diagonals through P and Q and $\dot{y}=0$ on

78

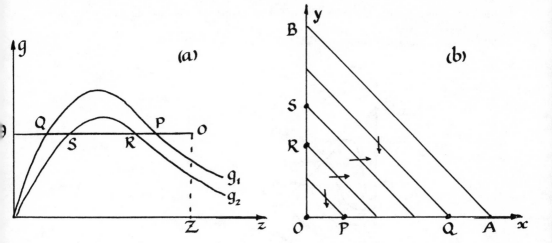

those through R and S. Moreover we can put arrows on the line segments

for we see by part (a) of the figure that $g_2 < \theta$ at P and so $\dot{y} < 0$ on

the diagonal through P in part (b). The axes x=0 and y=0 are solutions

of the equations and we can put arrows on the segments by glancing at

part (a) of the figure. The other diagonals are not isoclines but we notice

that on $x+y = Z-\zeta$

$$\frac{dy}{dx} = K \frac{y}{x} : \quad K = \frac{g_2(\zeta)-\theta}{g_1(\zeta)-\theta} ,$$

that is the tangent of the angle that is the direction of the solution is a

constant multiple of (y/x), the tangent of the direction from (x,y) to the

origin. K = 1 on AB and zero or infinite on the diagonals through the

steady states, and we can see that it varies roughly in this fashion:

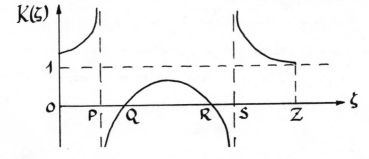

Thus in the trapezoidal region between the diagonals through R and S the flow must be northeasterly and the slope of the paths increase to a maximum and then decrease again to zero.

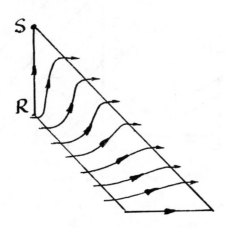

The direction of the solutions on AB is straight to the origin since K = 1 on AB. If this is done for each section a portrait emerges in which there will be south-south-easterly and east-south-easterly trajectories connecting R with P and S with Q in their respective trapezoids.

Thus we have already a general impression of the solution and see that RP will be a separatrix between the regions of attraction of O and Q the two stable steady states. We also see where computation may be difficult. For example, if it is important to be accurate in the neighbourhood of R the rapid change of direction may give trouble.

80

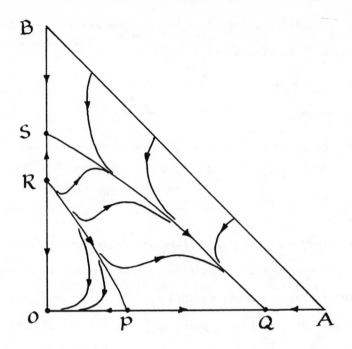

All this has been done without any computation. It can be confirmed (or perhaps eased over a sticking point) by linearization. Eqns. (4.76) and (4.77) can be linearized about any steady state in the form

$$\dot{\xi} = F\xi - G\eta$$
$$\dot{\eta} = -H\xi + K\eta$$

(4.78)

where

$$x = x_s + \xi, \; y = y_s + \eta$$

(4.79)

and

$$F = g_1(Z-x_s-y_s) - \Theta - x_s g_1'(Z-x_s-y_s)$$

$$G = x_s g_1'(Z-x_s-y_s)$$

(4.80)

$$H = y_s g_2'(Z-x_s-y_s)$$

$$K = g_2(Z-x_s-y_s) - \Theta - y_s g_2'(Z-x_s-y_s)$$

81

are constants since they are evaluated at the steady state under scrutiny. Because of the equations satisfied at the various steady states, the constants $F,\ldots K$ simplify as follows

Type of Steady State	F	G	H	K
0	$g_1 - \theta$	0	0	$g_2 - \theta$
P,Q	$-G$	G	0	$g_2 - \theta$
R,S	$g_1 - \theta$	0	H	$-H$
T,U	$-G$	G	H	$-H$

The next table shows what the eigenvalues and eigenvectors of the linearization are and what the conditions are that give the steady state its character. These characters are denoted by SN for stable node, UN for unstable node and SP for saddle point. The suffix on SP tells which eigenvector is tangent to the unique incoming trajectory. In the case of T and U one eigenvalue is always zero and the stability or instability of a point is governed by the sign of the other; these are denoted by SS and SU for semi-stable and semi-unstable.

From this table we can read the character of the steady states in the example OPRSQ above. D is a stable node, P a saddle point (SP_2), Q another stable node, R an unstable node and S an SP_2.

Steady State	Eigenvalues and vectors $\lambda_1, \underset{\sim}{v}_1$	$\lambda_2, \underset{\sim}{v}$	Character and conditions
0	$F, \begin{bmatrix} 1 \\ 0 \end{bmatrix}$	$K, \begin{bmatrix} 0 \\ 1 \end{bmatrix}$	SN, g_1, $g_2 < \Theta$ $SP_1, g_1 < \Theta < g_2$ $SP_2, g_2 < \Theta < g_1$ UN, $\Theta < g_1$, g_2
P,Q	$-G, \begin{bmatrix} 1 \\ 0 \end{bmatrix}$	$K, \begin{bmatrix} -G \\ K+G \end{bmatrix}$	SN, $G > 0$, $g_2 < \Theta$ $SP_1, G > 0$, $g_2 > \Theta$ $SP_2, G < 0$, $g_2 < \Theta$ UN, $G < 0$, $g_2 > \Theta$
R,S	$F, \begin{bmatrix} F+H \\ -H \end{bmatrix}$	$-H, \begin{bmatrix} 0 \\ 1 \end{bmatrix}$	SN, $H > 0$, $g_1 < \Theta$ $SP_1, H < 0$, $g_1 < \Theta$ $SP_2, H > 0$, $g_1 > \Theta$ UN, $H < 0$, $g_1 > \Theta$
T,U	$0, \begin{bmatrix} -1 \\ 1 \end{bmatrix}$	$-G-H, \begin{bmatrix} G \\ H \end{bmatrix}$	SS, $G + H > 0$ SU, $G + H < 0$

4.5.2 Coarse numerical methods and their uses.

In the preceding section we saw that the partial differential equation for diffusion and reaction in a catalyst particle could be reduced to

$$\nabla^2 \xi + R(\xi) = 0 \text{ in } \Omega$$

$$\xi = 0 \text{ on } \partial\Omega. \tag{4.66 bis}$$

Let us suppose that Ω is either an infinite slab, an infinite cylinder or a sphere so that the equation becomes

$$\frac{1}{\rho^q} \frac{d}{d\rho} \left[\rho^q \frac{d\xi}{d\rho} \right] + R(\xi) = 0, \tag{4.81}$$

$$\xi = 0, \quad \rho = 1, \tag{4.82}$$

with $q = 0$, 1 or 2 and ρ is the fractional radius or distance from the central plane. For reasons that we will not go into here (cf. [14]), the form of $R(\xi)$ is:

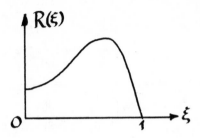

This is clearly a very nonlinear function but is positive so that the solution is a convex function of ρ. In the case of the slab we can solve the equation by quadrature but we wish to follow a route opened by Villadsen and Stewart [188] and use a one point collocation method. For fuller details of this problem Finlayson's monograph [66] or Villadsen and Michelsen's book [261] should be consulted.

Granted that the solution is a convex function of ρ and zero at $\rho=1$, a crude approximation to it would be $\xi(\rho) = \alpha(1-\rho^2)$ and its Laplacian is

$$-2(q+1)\alpha = -2(q+1)\xi(\rho)/(1-\rho^2).$$

Hence if α were chosen to satisfy the equation at a given point, say $\rho=\alpha$, the value of $\xi(\sigma)=\zeta$ would satisfy

$$\frac{2(q+1)}{1-\sigma^2}\zeta = R(\zeta) \qquad (4.83)$$

There is a full-blooded theory for the choice of collocation points such as σ, but we will give only the simplest motivation. If $q=2$ and $R(\xi) = \phi^2(1-\xi)$ is linear then eqn. (4.81) can be solved to give:

$$\xi(\rho) = 1 - \frac{\sinh\phi\rho}{\rho\sinh\phi}$$

If the same $R(\xi)$ is used in eqn. (4.83) with $q=2$,

$$\xi(\rho) = \frac{\phi^2(1-\rho^2)}{6+\phi^2(1-\sigma^2)}$$

These expressions should agree as far as possible and we find that they do

so for small ϕ if $\sigma^2 = 3/7$. A similar comparison for other cases gives

$\sigma^2 = (q+1)/(q+5)$, giving the equation

$$\frac{(q+1)(q+5)}{2}\,\zeta = R(\zeta) \tag{4.84}$$

for $\zeta = \xi(\sigma)$ and $\alpha = \zeta/(1-\sigma^2) = (q+5)\zeta/4$.

Returning now to the nonlinear form of $R(\xi)$ we have an immediate

graphical construction for ζ by drawing a line of the appropriate slope

from the origin. When we do this we see there will be

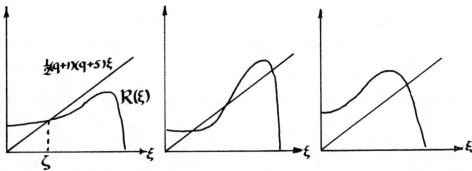

cases where the line will intersect more than once with the curve $R(\xi)$. We

are thus warned of the possibility of multiple solutions of the original

equation even though we acknowledge that the approximation is crude to a

degree. But we also note that if $R(\xi)/\xi$ is monotonic decreasing as ξ

increases from 0 to 1, then there can never be more than one intersection.

This suggests a criterion for uniqueness that might be usable in the

original partial differential equation. Suppose ξ_1 and ξ_2 are two

distinct solutions of $\nabla^2\xi + R(\xi) = 0$. By Green's theorems and the fact

that $\xi_1 = \xi_2 = 0$ and Ω

$$\iiint\limits_{\Omega} \{(\xi_1 \nabla^2 \xi_2 - \xi_2 \nabla^2 \xi_1)dV = 0$$

Hence substituting from the equations for the Laplacians,

$$0 = \iiint\limits_{\Omega} \{\xi_1 R(\xi_2) - \xi_2 R(\xi_1)\} \, dV$$

$$= \iiint\limits_{\Omega} \left[\frac{R(\xi_2)}{\xi_2} - \frac{R(\xi_1)}{\xi_1} \right] \xi_1 \xi_2 dV.$$

Now it can be shown (see [105]) that there is a maximal solution and if we choose ξ_1 to be this $\xi_2 \leq \xi_1$ everywhere. If then $R(\xi)/\xi$ is monotonic the integral is positive and can only be zero if $\xi_2 = \xi_1$. The monotonicity of $R(\xi)/\xi$ is thus a very general sufficient condition for uniqueness (cf. [122]).

4.5.3 The interaction of easier and more difficult problems.

In this final example we want to explore further aspects of the craft of seeing what can be learned about the model by getting a qualitative feel for the solution before any actual calculation is done. In particular the interplay of the various versions of a model will be emphasized and we shall see how appeal to an apparently more difficult problem can sometimes illuminate a simpler one. The physical system S is a vertical counter-current contacting column through which a gas flows upwards while a solid falls downwards through the gas flow and is removed at the bottom. A substance is brought in with the gas stream where its concentration at height x and time t is $c(x,t)$. It can be adsorbed on the solid where its concentration is denoted by $n(x,t)$. Both these concentrations are the amounts of A per unit volume of the phase and ε and $(1-\varepsilon)$ are the volume fractions of the two phases. Once adsorbed on the solid A can either be desorbed back into the gas or react to form a product B which is instantaneously desorbed. Because of this instantaneous desorption we have no need to consider an equation for the concentration of B simultaneously with those for the concentrations of A, but can determine this after the solution for A. The scheme is $A = A^* \to B$, where A^* denotes the absorbed form of A. The figure shows the notation and general scheme. We shall not dwell on the hypotheses nor on the reduction of the equations to dimensionless form since we are interested in the subsequent treatment of the equations. Suffice it to say that the rate of adsorption is $k_a(N-n)c$ where N is the saturation value of n, and the rates of desorption and reaction are $k_d n$ and $k_r n$ respectively. Then the usual balances give

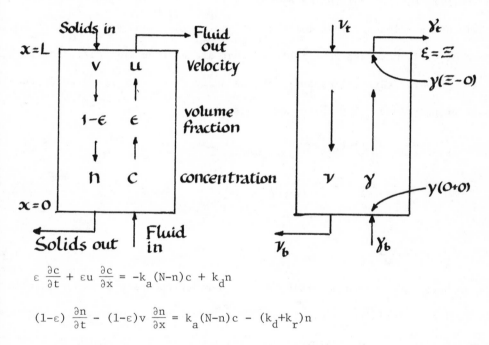

$$\varepsilon \frac{\partial c}{\partial t} + \varepsilon u \frac{\partial c}{\partial x} = -k_a(N-n)c + k_d n$$

$$(1-\varepsilon) \frac{\partial n}{\partial t} - (1-\varepsilon)v \frac{\partial n}{\partial x} = k_a(N-n)c - (k_d+k_r)n$$

These equations can be added to give

$$\frac{\partial}{\partial t} \{\varepsilon c+(1-\varepsilon)n\} + \frac{\partial}{\partial x} \{\varepsilon uc-(1-\varepsilon)vn\} = -k_r n$$

which would be a convenient form if n were known as a function of c.

They will be rendered dimensionless by the following transformations.

$$\gamma = Kc = k_a c/k_d, \quad \nu = n/N$$

$$\xi = k_r x/\varepsilon u, \quad \Xi = k_r L/\varepsilon u, \quad \tau = k_r t/\varepsilon$$

$$\lambda = Nk_a/k_r, \quad \mu = Nk_a/k_d, \quad \alpha = \mu(1-\varepsilon)/\varepsilon, \quad \sigma = \alpha v/u$$

There is no need to swell on this reduction except to draw attention to the meaning of α and σ. $\mu = NK$ is the ratio of adsorbed to fluid phase concentrations in the limit of dilute equilibrium and, in fact, is an upper bound of this ratio. In α this is multiplied by the volume ratio of the two phases and so is the ratio of the amounts that can be held in each. In σ this is further multiplied by the ratio of the velocities and so is the ratio of the fluxes. This interpretation should already suggest that the case of $\sigma > 1$ may have some features that are particularly different from $\sigma < 1$ since the carrying capacity of the solid stream is then greater than that of the fluid. The three equations are:

$$\frac{\partial \gamma}{\partial \tau} + \frac{\partial \gamma}{\partial \xi} = -\lambda\{(1-\nu)\gamma-\nu\} \tag{4.85}$$

$$\alpha \frac{\partial \nu}{\partial \tau} - \sigma \frac{\partial \nu}{\partial \xi} = \lambda\{(1-\nu)\gamma-\nu\} - \mu\nu \tag{4.86}$$

and

$$\frac{\partial}{\partial \tau} \{\gamma+\alpha\nu\} + \frac{\partial}{\partial \xi} \{\gamma - \sigma\nu\} + \mu\nu = 0 \tag{4.87}$$

We have four models according as we consider the rate of adsorption to be finite or infinite (i.e. λ finite or $\lambda \to \infty$) and according as we conside the transient or steady state. If $\lambda \to \infty$ the only way in which the right hand side of eqn. (4.85) can remain finite is for the equilibrium relation $(1-\nu)\gamma-\nu = 0$ to obtain. Thus

$$\nu = \gamma/(1+\gamma)$$

and, substituting in eqn. (4.87)

$$\left\{1 + \frac{\alpha}{(1+\gamma)^2}\right\} \frac{\partial \gamma}{\partial \tau} + \left\{1 - \frac{\sigma}{(1+\gamma)^2}\right\} \frac{\partial \gamma}{\partial \xi} + \frac{\mu\gamma}{1+\gamma} = 0 \qquad (4.88)$$

In the steady state

$$\frac{d\gamma}{d\xi} = -\lambda\{(1-\nu)\gamma - \nu\} \qquad (4.89)$$

$$\frac{d\nu}{d\xi} = -\lambda\{(1-\nu)\gamma - \nu\} + \mu\nu \qquad (4.90)$$

or

$$\left\{1 - \frac{\sigma}{(1+\gamma)^2}\right\} \frac{d\gamma}{d\xi} + \frac{\mu\gamma}{1+\gamma} = 0 \qquad (4.91)$$

The four models are:

Model	Non-equilibrium λ finite	Equilibrium $\lambda \to \infty$
Transient	4.85, 86, 92–95 :Σ	4.88, 92, 94–5 :Σ_e
Steady-state	4.89, 90, 94–5 :Σ_s	4.91, 94–5 :Σ_{es}

For simplicity we shall consider only constant inlet and boundary conditions. The initial conditions are

$$\gamma(\xi,0) = \gamma_o \qquad (4.92)$$

$$\nu(\xi,0) = \nu_o. \qquad (4.93)$$

The inlet conditions specify γ at the bottom and just outside the column, whereas ν is specified just above the top. We write these as

$$\gamma(0-0,\tau) = \gamma_b \qquad (4.94)$$

$$\nu(\Xi+0,\tau) = \nu_t. \qquad (4.95)$$

This is physically correct but we should sense that there may be a problem since Σ_{es} would appear to be overdetermined. The resolution of this difficulty will appear.

We are interested in how these models can be used to illuminate one another and how the scope of the solutions can be understood and their general form obtained without actually computing anything in detail. Obviously Σ is the most difficult model to crack and though there are well-known methods of treating such equations it would be foolhardy to do so without preliminary consideration. We already have the limiting case $\lambda \to \infty$ under consideration and the solution for large λ will presumably be a slight blurring of this limiting case. The limit $\lambda \to \infty$ is not of the same interest for in this case the adsorption equilibrium is so slow that the two streams do not exchange matter at all. Σ_s and Σ_{es} should be obtainable by letting time run out to infinity, but let us start with the simplest model Σ_{es}.

The equation for Σ_{es} is separable and gives

$$\mu\xi = \int_{\gamma(\xi)}^{\gamma(0)} \left\{ \frac{1+\gamma}{\gamma} - \frac{\sigma}{\gamma(1+\gamma)} \right\} d\gamma \qquad (4.96)$$

This can be easily integrated but, for the moment, let us examine it without explicitly integrating, since a similar, but not so easily evaluated integral, might arise in a more complex problem. If $\sigma < 1$ the integrand is always positive and, since the integrand behaves like $(1-\sigma)/\gamma$ for small values of γ, it is clear that $\xi \to \infty$ as $\gamma(\xi) \to 0$ and a solution can be found for any value of Ξ. In fact all that is needed is to take a segment of

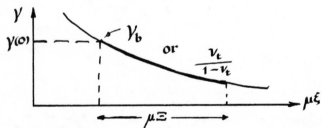

length $\mu \Xi$ of the curve. If we set $\gamma(0) = \gamma_b$ this defines the segment. But we could equally well suppose that the value of $\gamma(\Xi)$ is equilibrium

90

with ν_t and this would fix the segment by fixing the value of γ at the top. Having both conditions is physically sensible, but, mathematically, it is too much and, except in very special circumstances, is going to be contradictory. We might reconcile the physical necessity of specifying both γ_b and γ_t with the mathematical impossibility of generally satisfying both $\gamma(0)=\gamma_b$, $\gamma(\Xi)=\nu_t/(1-\nu_t)$, by saying that there must be a discontinuity at one end or the other. Thus the solid as soon as it contacts the gas at $\xi=\Xi-0$ might (thanks to the fact that λ is infinite) instantaneously take up the equilibrium value $\nu(\Xi)=\gamma(\Xi)/\{1+\gamma(\Xi)\}$. In order for there to be no continuous build-up of matter at the top the concentration γ must also have a discontinuity from $\gamma(\Xi)=\gamma(\Xi-0)$ to $\gamma_t=\gamma(\Xi+0)$. A balance round the plane $\xi=\Xi$ shows that the condition for no accumulation in the plane is

$$\gamma(\Xi) - \gamma_t = \sigma\{\nu(\Xi)-\nu_t\}$$

Thus, if the discontinuity is at the top, the concentration in the emerging fluid is not $\gamma(\Xi)$, but

$$\gamma_t = \gamma(\Xi) - \sigma \left\{\frac{\gamma(\Xi)}{1+\gamma(\Xi)} - \nu_t\right\}. \qquad (4.97)$$

But why should this happen at the top? From the mathematics it could equally well happen at the bottom or, for that matter, at both ends. We could argue physically that the case $\sigma<1$ must be continuous with $\sigma=0$ and that in this case the bed of solid is fixed and only the condition at the bottom can be insisted upon. But to see this mathematically we have to look at one or other of the harder problems Σ_e or Σ_s to get the needed insight into the allegedly simpler Σ_{es}. Let us pick Σ_e leaving the reader to look into Σ_s, since more will be said about this later.

Σ_e is a single first-order quasilinear partial differential equation which can be readily solved by the method of characteristics. In fact the characteristic equations of eqn. (4.88) are

$$\frac{d\tau}{ds} = 1 + \frac{\alpha}{(1+\gamma)^2} \tag{4.98}$$

$$\frac{d\xi}{ds} = 1 - \frac{\sigma}{(1+\gamma)^2} \tag{4.99}$$

$$\frac{d\gamma}{ds} = -\frac{\mu\gamma}{1+\gamma} \tag{4.100}$$

Again these are easy enough to integrate explicitly but, having in mind that we want to adopt tactics that would work with more complicated equations, we will lightly avoid this. We first notice that γ decreases monotonically along a characteristic so that it might be taken as the parametric variable along the characteristic. Clearly τ always increases along the characteristic and, since $\sigma<1$, so does ξ. Thus the characteristics always have a positive slope. In the ξ,τ plane we are interested in the strip $\tau>0$, $0 < \xi < \Xi$ and this can be covered by characteristics emanating from the initial interval $\tau=0$, $0 < \xi < \Xi$ where $\gamma=\gamma_0$ and the inlet at the bottom $\xi=0$, $\tau<0$ where $\gamma=\gamma_b$. The characteristics emanating from points on

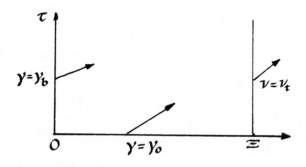

the top, $\xi=\Xi$, are directed outwith the region of interest so that conditions specified at the top can have no influence there. The discontinuity described by eqn. (4.97) must therefore be present at the top of the column. This is entirely in keeping with our feelings about the meaning of $\sigma<1$, for the fact that the carrying capacity of the solids is less than that of the fluid means that the fluid stream can "blow out" the influence of the solids. Thus v_t has no influence at all on the solution inside the reactor, but it does have an influence, through eqn. (4.97) on γ_t, the product of the reactor. This resolution satisfies both our mathematical and physical expectations.

To complete the case $\sigma<1$ let us note that the slope of characteristic

$$\left[\frac{d\tau}{d\xi}\right]_c = \frac{(1+\gamma)^2+\alpha}{(1+\gamma)^2-\sigma} \tag{4.101}$$

is positive and increases as γ decreases going from near 1 when γ is large to $(1+\alpha)/(1-\sigma)$ as $\gamma\to0$ (see section (a) of figures below). If $\gamma_b<\gamma_o$

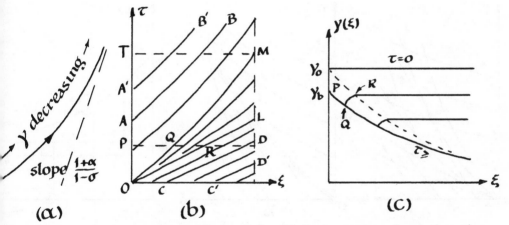

(α) (b) (c)

the characteristics emanating from $\xi=0$ all have greater slope than those that emanate from $\tau=0$. There is thus no tendency for the two sets of characteristics to overlap one another; the one set is vertically parallel like AB and A'B' and the other horizontally parallel like CD and C'D'. Moreover at the origin all concentrations between γ_o and γ_b may be thought to be present. The characteristics corresponding to these therefore fan out and fill in the region LOM. Taking sections at constant values of τ we see that the profile of the steady state is established in the length of reactor up to the characteristic OM then there is a

93

transition region between OM and OL beyond which the initial constant state is decaying and being pushed out of the reactor. The final profile (PS in Section C of the figure) is established in the time T).

On the other hand if $\gamma_o < \gamma_b$ the characteristics from $\xi = 0$ have a smaller slope than those from $\tau = 0$ and will overlap one another and all the characteristics emanating from the origin (section (a) of figure below). This is intolerable as it would imply that three different concentrations were present at one point. The resolution of this is to introduce a shock or discontinuity. The speed of a shock, w, is such that its movement durin

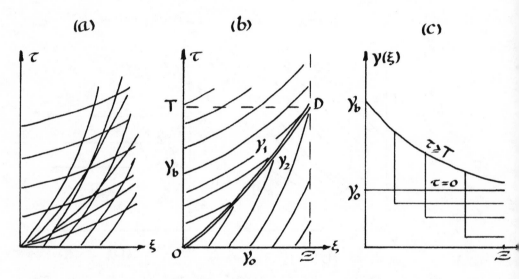

a time interval δt, namely $w\delta t$, accounts for the net access of material. Thus if c_1, n_1 are the values of the concentrations just below and c_2, n_2 those just above the discontinuity

$$w\delta t [\varepsilon (c_1 - c_2) + (1-\varepsilon)(n_1 - n_2)]\delta t$$

$$= [u\varepsilon(c_1-c_2)-v(1-\varepsilon)(n_1-n_2)]\delta t$$

or

$$\left[\frac{d\xi}{d\tau}\right]_d = \frac{w}{u} = \frac{[\gamma] - \sigma[\nu]}{[\gamma] + \alpha[\nu]}$$

94

where $[\gamma] = \gamma_1 - \gamma_2$ and $[\nu] = \nu_1 - \nu_2 = [\gamma]/(1+\gamma_1)(1+\gamma_2)$.

Thus the slope of a shock line in the ξ,τ-plane is

$$\zeta_{12} = \left[\frac{d\tau}{d\xi}\right]_d = \frac{(1+\gamma_1)(1+\gamma_2)+\alpha}{(1+\gamma_1)(1+\gamma_2)-\sigma} \tag{4.102}$$

This equation is confirmed by the fact that when the discontinuity vanishes, i.e. $\gamma_1 = \gamma_2 = \gamma$, we have the slope of the characteristic. We also note that ζ_{12} is certainly positive as long as $\sigma < 1$. If ξ_s,τ_s is a point on the shock line with γ_1 behind and γ_2 before we can calculate the path of the shock as follows. γ_1 must correspond to a distance ξ_s along a characteristic that emanates from $\xi = 0$ where $\gamma = \gamma_b$. Thus, dividing eqn. (4.98) by eqn. (4.100) and separating variables,

$$\mu\xi_s = \int_{\gamma_1}^{\gamma_b} \left[\frac{1+\gamma}{\gamma} - \frac{\sigma}{\gamma(1+\gamma)}\right] d\gamma \tag{4.103}$$

Similarly, γ_2 is the value of γ on a characteristic emanating from $\tau = 0$ where $\gamma = \gamma$; thus from eqns. (4.99) and (4.100)

$$\mu\tau_s = \int_{\gamma_2}^{\gamma_0} \left[\frac{1+\gamma}{\gamma} + \frac{\alpha}{\gamma(1+\gamma)}\right] d\gamma \tag{4.104}$$

Moreover $d\tau_s/d\xi_s = \zeta_{12}$ is given by eqn. (4.102) so that these equations, 4.102, 103 and 104, provide a way of calculating the path of the shock. We are not concerned here with the question of how these equations can be solved, but the solution must clearly give a path such as OD above. The final steady state solution is fully developed behind the shock while the initial state decays and is pushed out in front (i.e. above) it. Again the steady state is established in a finite, calculable time T.

Let us now consider the case $\sigma > 1$ and return at first to the steady state equation of Σ_{es}. The solution is again given by eqn. (4.96) for sufficiently small values of ξ. But we notice that when γ drops to $(\sigma^{1/2}-1)$ the integrand vanishes and subsequently becomes negative. This means that as $\gamma(\xi)$ continues to fall the value of ξ decreases instead of increasing, giving the solution shown in the figure. But this is physically impossible since it gives two values of γ for some ξ and cannot give

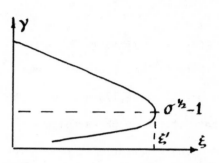

a solution if $\Xi > \xi'$. This suggests that there must be discontinuities in the solution and to see where these arise we turn this time to Σ_s.

Σ_s is a pair of ordinary differential equations, (4.89) and (4.90), whose right hand sides are functions only of γ and ν. We can combine them into a single differential equation which will give the relation between ν and γ at any point by dividing the one by the other

$$\frac{d\nu}{d\gamma} = \frac{1}{\sigma} - \frac{\mu}{\lambda\sigma} \frac{\nu}{(1-\nu)\gamma-\nu} \qquad (4.105$$

The only critical point is the origin, which is a saddle-point having an entering trajectory of slope η_- and departing trajectory of slope η_+ where

$$\eta_\pm = \frac{1}{2\sigma}\left[1 + \sigma + (\mu/\lambda)\pm \sqrt{\{1+\sigma+(\mu/\lambda)\}^2-4\sigma.}\right]$$

In fact the isoclines $d\nu/d\gamma = \eta$, are

$$\nu = \frac{(\lambda/\mu)(1-\sigma\eta)\gamma}{1 + (\lambda/\mu)(1-\sigma\eta)(1+\gamma)} = \frac{\beta\gamma}{1+\beta\gamma} \qquad (4.106$$

where

$$\beta = \frac{\lambda(1-\sigma\eta)}{\mu+\lambda(1-\sigma\eta)} \quad .$$

In particular there are the two loci of horizontal ($\eta=0$) and vertical ($\eta=\infty$) directions and we note that $\eta_+ > \beta_\infty > \beta_0 > \eta_-$. Thus the phase plane looks as below where OA and OB are the trajectories through the saddle point O and OC and OD the isoclines of verticality and horizontality.

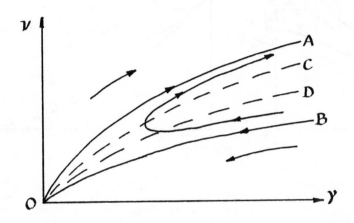

To solve eqn. 4.105 subject to the conditions $\gamma(0)=\gamma_b, \nu(\Xi) = \nu_t$ we have to find a path in the phase plane between the vertical line $\gamma=\gamma_b$ and the horizontal $\nu=\nu_t$. Not any such trajectory will do but if the reactor is of length Ξ we must find the path for which eqn. (4.89) gives

$$\lambda \Xi = \int_{\gamma_b}^{\nu_t} \frac{d\gamma}{(1-\nu)\gamma-\nu} \ .$$

This integral requires a bit of interpretation when the path crosses $\nu=\gamma/(1+\gamma)$ since the integrand there becomes infinite. If $ds^2 = d\gamma^2 + d\nu^2$ is the path length along the trajectory and $f = f(\gamma,\nu) = (1-\nu)\gamma-\nu$, then the integral could be written

$$\lambda \Xi = \int \frac{\sigma ds}{\{\sigma^2 f^2 + (f-\mu\nu)^2\}^{1/2}} \ , \qquad (4.107)$$

a form which is unexceptionable. For example, if the point (γ_b,ν_t) lies between the arms OA and OB there are a number of possible paths: P itself of length zero; QR corresponding to a short length; ST to a

longer reactor and UV to a very long one; the two segments XO and OY

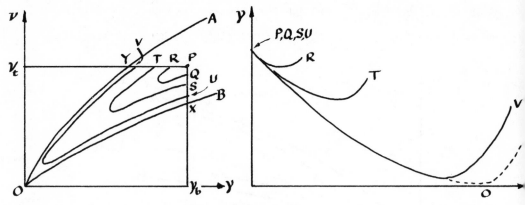

of OB and OA would give an infinite reactor since it takes an infinite

length to get into and out of the origin. If the point (γ_b, ν_t) lies

beneath OB the situation is somewhat different as shown below. Again

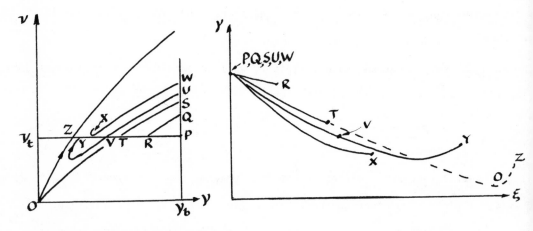

there is the path of zero length P, a short path QR and longer one s,

ST and UV. These are monotonic until WX, but to fit a reactor longer

than the Ξ corresponding to the path WX we must go back to a curve such

as UXY which follows an earlier solution UV but dips below $\nu = \nu_t$ and

comes up again. STOZ is the ultimate solution of this kind since it takes

an infinite length to get into and out of O. (Note that paths such as ZA do not generate solutions since they go from the horizontal $\nu=\nu_t$ to the vertical $\gamma=\gamma_t$ and not vice-versa).

How does the situation look in the limit $\lambda\to\infty$? This will take us back to Σ_{es} where we foresaw certain difficulties in getting any solution at all if $\Xi>\xi'$. We have no space to classify exhaustively all the possibilities (this has been done by Aris and Viswanathan [18]) so will take only one case. As $\lambda\to\infty$ all the isoclines in the figure on p. 97 close down on the one curve $\nu=\gamma/(1+\gamma)$ and this curve, Γ, is a solution of the equation. This is not surprising for $\lambda\to\infty$ means that adsorption equilibrium is instantaneous so that we should expect the equilibrium relationship to obtain. At any point not on Γ the limit $\lambda\to\infty$ in eqn. 4.105 gives $d\nu/d\gamma = 1/\sigma$, i.e. the solution is a straight line. Such a solution corresponds to a discontinuity for with λ infinite any path integral such as eqn. 4.107 shows that a segment of such a line is traversed in zero length. If $\sigma>1$ there is a point, C, on the curve Γ of slope $1/\sigma$, in fact $\gamma=\sigma^{1/2}-1$ at this point. Below it the curve is traversed upwards and above the trajectory is down-wards. The phase plane is :-

where all the straight line segments all correspond to discontinuities.

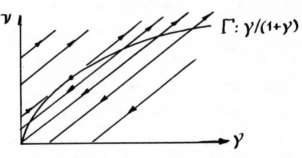

Consider a pair of boundary conditions γ_b, ν_t that give a point close under the upper part of Γ as shown (in fact $\gamma_b>\sigma-1$, $1-\sigma^{-1/2}<\nu_t<\gamma_b/(1+\gamma_b)$ as the longer analysis of Aris and Viswanathan shows). The possible ways of getting from the vertical $\gamma=\gamma_b$ to the horizontal $\nu=\nu_t$ are shown numbered in order of increasing reactor length. 1 is trivial corresponding to a reactor of zero length. In 2 the reactor is so short that the concentration does not fall very much in the reactor, it is continuous at the bottom but has a discontinuity at the top. 3 is the segment of Γ between $\gamma=\gamma_b$ and $\nu=\nu_t$ and corresponds to the only length Ξ for which there is a continuous solution in the closed interval $0 < \xi < \Xi$. For a slightly longer reactor, 4, we can go below the horizontal and come back to it with a discontinuity at the exit and this can be done until we reach the point C. This point, where $\gamma(\Xi)=\sigma^{1/2}-1$, corresponds to a reactor of length ξ' as in the figure of p. 96. How do we get a one-valued

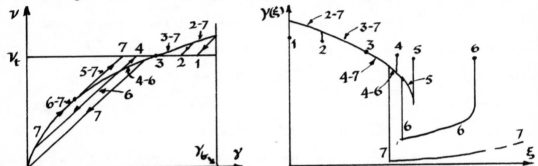

solution if $\Xi>\xi$? The only way is to introduce a discontinuity in the middle of the solution, following the curve Γ downwards to a point short of C, then taking a chord, such as 6, down to a point below C, following Γ back up to C and taking a jump to $\nu=\nu_t$. For all such solutions $\gamma(\Xi) = \sigma^{1/2}-1$. This type of solution can be fitted into any reactor that is long ($\Xi>\xi$), for a path such as 7 can be made very long since ξ moves rapidly with γ in the neighbourhood of the origin. Without doing any real calculation we have seen how the solution must lie.

Finally we may ask how this links up with the transient solution of Σ_e. Returning to the characteristic equations 4.98, 99 and 100 and taking decreasing γ as the parameter along the path of the characteristic we see that the solution of

100

$$-\mu \frac{d\xi}{d\gamma} = \frac{1+\gamma}{\gamma} - \frac{\sigma}{\gamma(1+\gamma)}$$

and

$$-\mu \frac{d\tau}{d\gamma} = \frac{1+\gamma}{\gamma} + \frac{\alpha}{\gamma(1+\gamma)}$$

is a curve of positive slope for $\gamma > \sigma^{1/2}-1$, which turns back on itself at $\gamma = \sigma^{1/2}-1$ and then has a negative slope ultimately decreasing to $-(\sigma-1)/(\alpha+1)$. We could in fact solve these equations once and for all and cut out a template for any pair of values of α and σ, graduating its edge in γ. It would look like this:

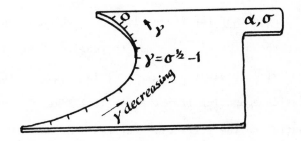

By laying such a template with γ_o at ξ_o, τ_o we can draw the character-istic emanating from such a point. For example, the characteristics emanating from $\xi=0$ can be drawn by placing γ_b on the vertical axis drawing along the template and then moving it upwards to give a sequence of parallel curves. Clearly if the curve is taken beyond $\sigma^{1/2}-1$ these curves can intersect one another so that there is an even greater need for discon-tinuities than before.

Suppose that γ_b and ν_t are disposed in the way we have just consi-dered, that $\gamma_o=0$ and that the reactor is long ($\Xi > \xi$). Then the character-istics emanating from the axis $\tau=0$ are straight lines of slope $-(\sigma-1)/(\alpha+1)$ which will certainly intersect those emanating from $\xi=0$. The resolution of this is to introduce a shock as we did before. This works well up to A the point where AB, the characteristic though $\Xi-0$ meets

101

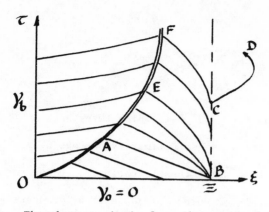

the shock. The characteristic from the vertical line, ξ-Ξ if drawn as

starting with $\gamma=\gamma_t=\nu_t/(1-\nu_t)$, the concentration γ in equilibrium with

ν_t, would point out of the region like CD, which is not permitted. However

we know from the study of Σ_{es} how to resolve this. There is a shock at

the top and $\gamma(\Xi)$ is always $\sigma^{1/2}-1$. This allows us to draw all the

characteristics emanating from the vertical like BE and CF--they are in

fact tangential to the vertical. Between BA and BE there is a fan of

characteristics corresponding to emanating concentrations between $\gamma_0=0$

and $\gamma(\Xi)=\sigma^{1/2}-1$. We now have all the characteristics and can start from O

and find the shock path that will satisfy eqn. 4.102. Cross-sections at a

typical sequence of points then show the development of the steady state

solution. In the next figure, $0,1,2,3,$ ∞ are the labels for an increasing

sequence of times. 0 is $\tau=0$ when $\gamma=0$; 1 is a time beneath the level of

A in the previous figure and there is still a midsection that is unaffected

by the inlet conditions either at the bottom or top; 2 is a time between the

levels of A and E before the top profile has assumed its final form, 3

is a time above E when all that remains for the discontinuity to move

forward to its final position, ∞.

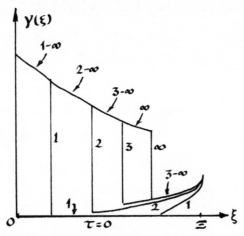

Thus we have built a framework within which the computation of the difficult case Σ can proceed and this framework can be made complete by exploring the whole range of possible dispositions of γ_b, ν_t and γ_o. It is interesting to note that though the internal steady state in Σ_e is only approached asymptotically as the internal shock moves into its final position, the output of the reactor reaches its final value instantaneously since $\gamma(\Xi,\tau)=\sigma^{1/2}-1$ for all τ. This is further discussed in [13].

5 How should a model be evaluated?

'Models are undeniably beautiful, and a man may justly be
proud to be seen in their company. But they may have their
hidden vices. The question is, after all, not only whether
they are good to look at, but whether we can live happily
with them.'

A. Kaplan. The Conduct of Inquiry.

5.1 Effective presentation of a model.

A mathematical model and the results that flow from its analysis deserve to
be presented effectively. There is nothing meretricious about this, it is
merely common-sense. If the model is worth studying and its analysis
illuminates the system then it should be presented in such a way that its
intrinsic merits or contributed understanding can be quickly grasped and
fairly assessed. One of the virtues of a model is that it can be studied
more or less comprehensively and with a little care can often ensure that
every representative case has been studied. It then becomes a question of
presenting it as effectively as possible. The phase plane is a good example
of this for by showing a sufficient number of trajectories it allows the eye
to visualize all possible solutions. Nor does the number of trajectories
have to be large to do this well. The limitation of such phase portraits is
that they are difficult to draw in more than two dimensions and one has to
make do with a few two-dimensional projections. Certainly coloured holo-
graphy would allow for four-dimensional presentations, but this is clearly
beyond the common reach; stereoscopic diagrams are possible but in their

usual form they require a decoupling of the eyes that not everyone can manage (cf. [157]). Nor is the artistic virtuosity of such a journal as Scientific American easily attained, though how effective such graphic art can be is seen in such articles as Zeeman's survey of catastrophe theory [197] (cf. also [255]). Even so the cusp is the highest canonical catastrophe that can be shown in its entirety; the catastrophe set of the swallowtail can be fully pictured, but the butterfly and other canonical forms have to be shown in section. Nevertheless part of the appeal of catastrophe theory is that it permits a more synoptic view of a greater range of behaviour than had been otherwise cultivated. (The literature of catastrophe theory grows apace: cf. [166,177,197,204,205,225,228,234,235, 250-2,255]).

The stirred tank reactor system Σ_6 is an example that can be well presented as topologically equivalent to a cusp catastrophe. Eqns. (4.11) and (4.13) for the steady state with a first order irreversible reaction reduce to eqn. (4.16), namely

$$(v - \bar{v})/\zeta = z(v;\alpha) \qquad (5.1)$$

where

$$z(v;\alpha) = \frac{\alpha e^{-1/v}}{1 + \alpha e^{-1/v}} \quad . \qquad (5.2)$$

When v_s, the steady state temperature is found, both sides of the equation are equal to the steady state reaction rate. So if we let

$$Z = z(v_s;\alpha) = Z(\alpha,\zeta, \bar{v}) \qquad (5.3)$$

we see that Z is a function of three parameters: α, the Damköhler number or intensity of reaction; ζ, a combination of heat of reaction and heat removal rate; $\bar{v}$, a mean feed/coolant temperature. If α is held constant the surface $Z(\alpha,\zeta, \bar{v})$ as a function of ζ and $\bar{v}$ takes on the form of a cusp catastrophe as has been known for sixty years. When the cooling capacity is very large $h \to 0$ and $\zeta \to 0$; when the operation is adiabatic $h \to 0$ and $\zeta \to \beta$. The possible modes of intersection are shown in the figure.

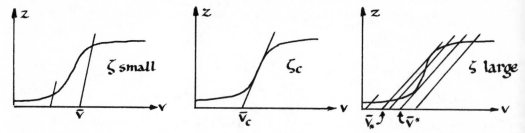

When ζ is small the straight line, which is the left hand side of the equation, is steeper than any part of the z-curve and the intersection is unique whatever the value $\bar{v}$. If ζ has the critical value ζ_c that gives the line the slope of the inflection point of the z-curve, the solution is still unique but the tip of the cusp is attained when $\bar{v}_c$ makes the line go through the inflection point. For $\zeta > \zeta_c$ then, for $\bar{v}_*(\zeta) < \bar{v} < \bar{v}^*(\zeta)$, there are three intersections.

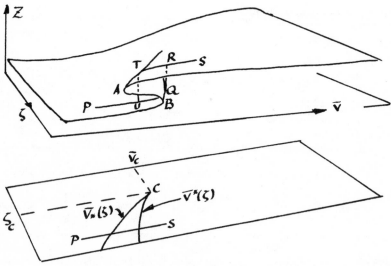

The lower part of the surface (left of CB) corresponds to a poor rate of reaction, where as the upper right (right of CA) represents a much higher reaction rate. The branches CA and CB of the catastrophe set might be

106

called the extinction and ignition catastrophes respectively, for if ζ is constant and $\bar{v}$ is increased from P to S and decreased back again, ignition takes place in a jump from Q to R and extinction in the drop from T to U.

The parameter α has been kept constant in this presentation and as it varies the cusp moves. We shall study this in some detail in the next section and it suffices to remark here that, though the cusp moves, no new features come to light (cf. Regenass and Aris [152]). However, two parameters are affected by the variation of flow rate and as was remarked in Sec. 4.3 a rather different parametrization is appropriate. Using the variables and parameters of eqns. (4.29 and (4.30) we have

$$\Theta\dot{u} = -u + \alpha\Theta(1-u)\exp[v/(1+v/\gamma)] \tag{5.4}$$

$$\Theta\dot{v} = \Theta v_c - (1+\Theta)v + \Theta\alpha\beta(1-u)\exp[v/(1+v/\gamma)] \tag{5.5}$$

This is essentially the same form of equations as that studied by Uppal, Ray and Poore [184] in one of the landmark papers of the subject. Following them, and for reasons we will not digress to here, we let $\gamma\to\infty$ and take $v_c=0$. There are still three parameters and one must be kept constant if we are to draw surfaces in three dimensions. Let β be fixed for the whole figure and plot the steady state reaction rate Z as a function of α and Θ. Such a surface is shown below and it is clearly somewhat different from the simple cusp surface having a Rehoboam-like finger protruding from the top of the wave. The catastrophe set is the hook-shaped region shown on the plane beneath.

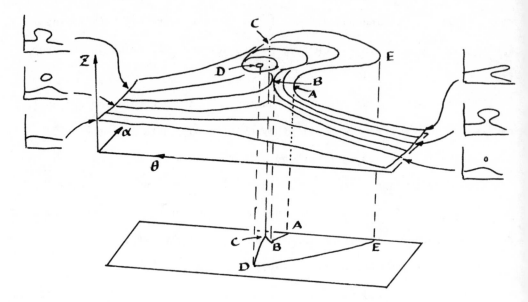

This is one way of presenting a small part of the results of Uppal, Ray and Poore but they have done vastly more than this by applying Hopf bifurcation to eqns. (5.4) and (5.5). By this means they have been able to direct their computations and discover all the possible types of behaviour of the system. These are shown in the next figure. Although the steady state behaviour is sufficiently demarcated by the hook-like catastrophe set (heavy line), within which there are three steady states and outside of which there is uniqueness, the dynamic behaviour is much more complicated. Nine different types of phase portrait can be discerned each corresponding to a different part of the α,θ plane. Thus if (α,θ) lies in region B there is a single steady state, but it is unstable and is surrounded by a stable limit cycle. Often a figure such as this will have to be distorted to show the detail, but it then becomes a map to guide the viewer into an accurately drawn figure.

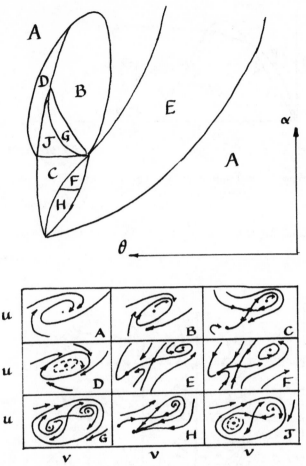

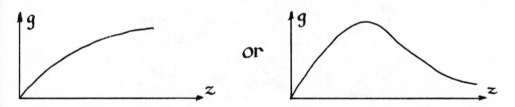

A slightly different form of comprehensive presentation can be given for the system we studied in Sec. 4.5.1. If the growth curves of the two organisms are either of the following forms:

or

and the label 1 is always used for the curve that is ultimately the higher, i.e. $g_1(z) \geq g_2(z)$ for large z, then we draw both curves in the same plane and put in the point (Z, θ). This is called 0 and reading from right to

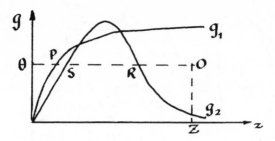

left the intersection with g_1 are first P then Q, if it is needed (it is not in the diagram above) and those with g_2 are first R then S. This gives a word (in the case above ORSP) and the corresponding phase portrait can be found in the gallery below (in the case above #21). This gallery together with the portrait of OPRSQ on p. 81 is complete except for confluence such O $\overline{RS}$

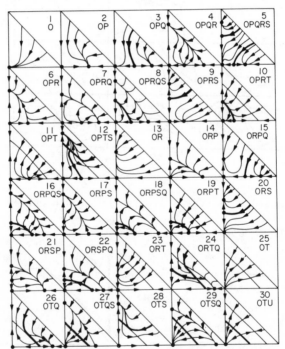

This confluence is not structurally stable for the slightest increase of will make $\overline{RS}$ disappear, while the slightest decrease will separate the confluence and give ORS. We have not been quite consistent in overlooking the structurally unstable however for we have included OT which arises from being exactly at the level of an intersection as part (b) of the next figure shows. We have even included OTU (shown in part (c) of the figure) which is "very unique" since it can only occur if the two intersections are at the same level.

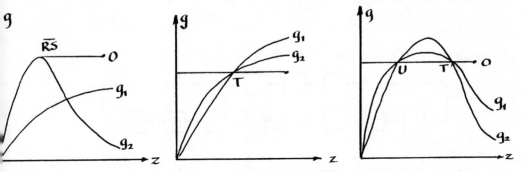

Two further examples of presentation may be given to illustrate the importance of bringing out a characteristic feature in as dramatic a fashion as possible. There are various ways of describing the residence time distribution in a flowing system. $P(t)$, the fraction of particles or molecules in the system at $t=0$ which have left by time t, is a monotone increasing function of t with $P(o)=0$, $P(\infty)=1$. It may be interpreted as the probability of a residence time being less than t. The expected residence time is $\mu = \int tP'(t)dt = \int\{1-P(t)\}dt$ and the variance of residence times $\sigma^2 = \int t\{1-P(t)\}dt-\mu^2$. These are the most obvious parameters and may be related to the position of the 'step' in the P-curve and to its steepness. An alternative description is $p(t)$ the probability density of residence times, i.e. $p(t)dt$ is the probability of a particle having a residence time in the interval $(t,t+dt)$. Clearly $p(t)dt = P(t+dt)-P(t)$, so that $p(t) = P'(t)$, $\mu = \int tp(t)dt$ and $\sigma^2 = \int t^2 p(t)dt-\mu^2$. The same information is present but this time the mean is the centre of gravity of a peak and the variance is a measure of its 'spread'. With an approximately Gaussian residence time distribution the one has little advantage over the other.

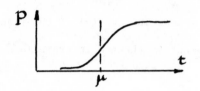 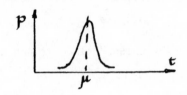

but where there is a rapid bypass the density function shows it up better. For example the system below would give a peak at the bypass residence time in the p-curve but this is rather lost in the P-curve.

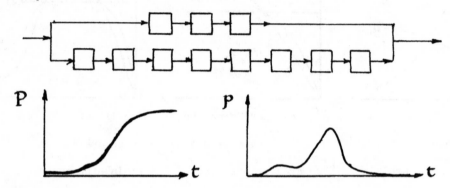

Shinnar has suggested the intensity function or escape probability density as another method of showing up the features of a residence time distribution. This is $\eta(t)dt$ the probability that a molecule of age t will leave the system during the interval $(t, t+dt)$. Since $1-P(t)$ is the fraction of molecules of age t in the system at time t, $p(t)dt = \{1-P(t)\}\eta(t)dt$ or

$$\eta(t) = \frac{p(t)}{1-P(t)} = \frac{d}{dt}\ln\{1-P(t)\}$$

For a single well-mixed compartment

$$p(t) = \frac{1}{\mu} e^{-t/\mu} \quad, \quad P(t) = 1-e^{-t/\mu} \quad, \quad \eta(t) = \frac{1}{\mu} \quad.$$

so that the departure of $\eta(t)$ from constancy is an indication of ill-mixedness. Some characteristic shapes are shown below, but for a full discussion of the intelligent use of residence time distributions the papers of Shinnar and his colleagues should be consulted [241-6].

112

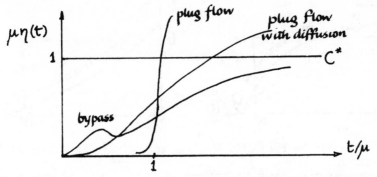

Another situation that lends itself to considered presentation arises
when two functions vary in opposite senses, the one increasing and the other
decreasing as functions of a controlling variable. Such a situation can
obtain when two pollutants behave contrarily with respect to an operating
variable

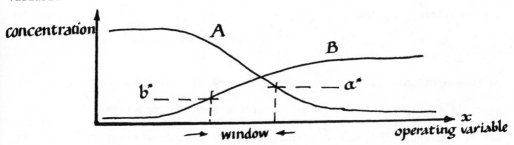

leaving a 'window' of acceptable conditions. If $a(n)$ and $b(n)$ are the
concentrations, with $a'(n)<0<b'(n)$, and acceptable levels of A and B
are $a*$ and $b*$. If $\alpha(a)$ and $\beta(b)$ are the inverse functions of $a(n)$
and $b(n)$ (i.e. $a(\alpha(a))\equiv a$, $b(\beta(b))\equiv b$), there will be an operating window if
$\beta(b*)<\alpha(a*)$. However this plot tells us little about the relative
behaviour of A and B and it is better to plot the curve $a=a(n),b=b(n)$
in the a,b-plane. The curve may be convex or concave toward the origin:

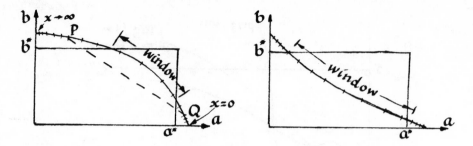

The window can again be read off, but now it becomes clear that it is easier to get away from both limitations when the curve is convex to the origin. On the other hand if the requirements can be met in the mean, they can more easily be met if the curve is concave to the origin for then the result of operating in slow oscillation between two such points as P and Q is represented a point on the chord PQ, which opens up a wider window for oscillatory operation.

5.2 Extensions of models.

In this section I want to raise a rather speculative question. When, if ever, is it profitable to extend consideration of a model into physically unreal regions? The answer could be, 'Never', since the physically correct regions should define itself and be closed. Thus, for example, with

$$V \frac{dc}{dt} = q(c_f - c) - VA_e^{-E/RT}c$$

$$VC_p \frac{dT}{dt} = qC_p(T_f - T) - h(T - T_{cf}) + (-\Delta H)VAe^{-E/RT}c$$

the region

$$0 \leq c \leq c_f, \quad 0 \leq T$$

is the physically appropriate region. Moreover we see that $(dc/dt) < 0$ on $c = c_f$, but $(dc/dt) > 0$ on $c = 0$; while $(dT/dt) > 0$ for $T = 0$ and (dT/dt) is negative for sufficiently large T.

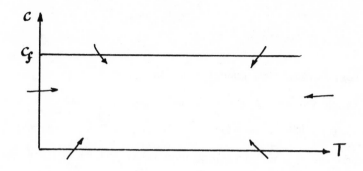

Thus the region of physical interest is closed and one could argue that attention should be confined to it. One cannot argue that attention can be confined to any sub-region on the grounds that only such a region is of practical significance unless it also is similarly closed. Thus D_1 would be an admissible confinement, but not D_2. But the equations themselves are indifferent toward physical reality, let alone practical significance,

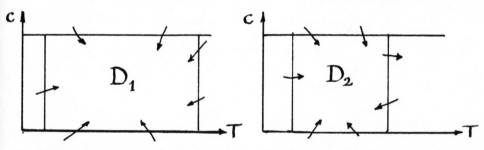

and obtain throughout the whole c,T plane. There is in this case the 'monstrous steep of Montmorency' at T=0 which we might expect to be a natural barrier.

But the enlargement of the variables beyond the bounds of physical reality is related to a similar extension of the parameters. In the above equations V, q, c_f, A, E, C_p, h, T_f and T_{cf} are all positive and only ΔH can change sign. In the non-dimensional form of the equations (cf. Sec. 4.3)

$$\frac{du}{d\tau} = 1 - u - \alpha u e^{-1/v} \qquad\qquad (5.6)$$

$$\frac{1}{\delta}\frac{dv}{d\tau} = \bar{v} - v + \alpha\zeta u e^{-1/v} \qquad\qquad (5.7)$$

the only parameter which can change sign is ζ.

However, there are at least two prima facie reasons why it might be desirable to extend the range of the parameters. First, the extension will allow us to see the system more comprehensively and understand its movements better. Second, it may suggest connections with other systems or even sugge new ones. Thus a stirred fermentor with growth by a logistical curve would have an equation

$$V \frac{dc}{dt} = -qc + kc(c_m - c)$$

with V, q, k and c_m positive. However allowing k to be negative gives the equation for a stirred tank with an autocatalytic reaction. Let us examine the steady state of the stirred tank in more detail.

The equations for u and v at steady state are

$$1 - u - \alpha u e^{-1/v} = 0,$$
$$\bar{v} - v + \alpha\zeta u e^{-1/v} = 0; \qquad\qquad (5.8)$$

these can be reduced to the single equation which has been cited as (5.1) but is repeated here:

$$\frac{1}{\zeta}(v - \bar{v}) = \frac{\alpha e^{-1/v}}{1 + \alpha e^{-1/v}} = Z(v;\alpha) \qquad\qquad (5.9)$$

The left hand side is a straight line, the right a curve with a discontinui at $v = 0$ and depending on only one parameter. There is a discontinuity a $v = 0$, since $z(0-0;\alpha) = 1$ but $z(0+0;\alpha) = 0$. The general form of the function is given on p. 117 for positive α and we see immediately that a arbitrary straight line can have 0, 1, 2 or 3 intersections. The forms fo

116

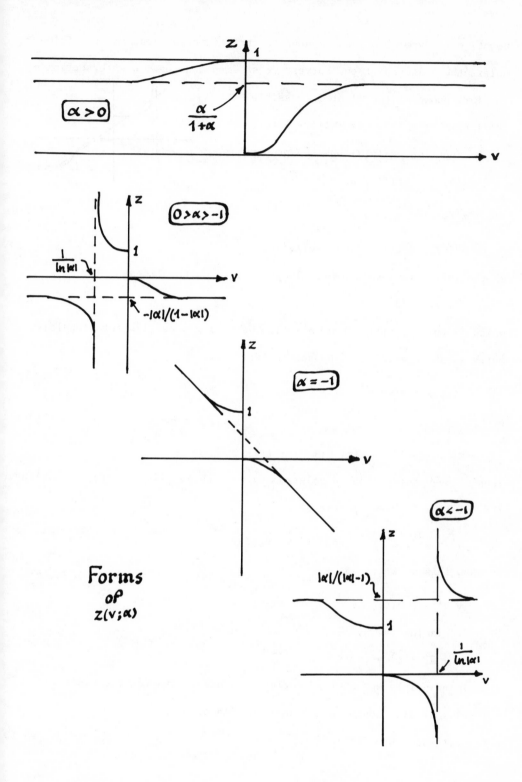

$\boxed{\alpha > 0}$

$\dfrac{\alpha}{1+\alpha}$

z

1

v

$\boxed{0 > \alpha > -1}$

z

1

$\dfrac{1}{\ln|\alpha|}$

v

$-|\alpha|/(1-|\alpha|)$

$\boxed{\alpha = -1}$

z

1

v

$\boxed{\alpha < -1}$

$|\alpha|/(|\alpha|-1)$

z

1

$\dfrac{1}{\ln|\alpha|}$

v

Forms
of
$z(v;\alpha)$

117

negative α are also shown and up to three intersections are possible

unless the branches of the curves are so disposed that an intersection of

the form shown on the right is possible. We

shall eliminate this possibility later.

The function $z(v;\alpha)$ has the properties

$$z(v;\alpha) \to \alpha/(1+\alpha) \text{ as } v \to \infty$$

$$z'(v;\alpha) = z(1-z)/v^2 \tag{5.10}$$

$$z''(v;\alpha) = z(1-z)(1-2z-2v)/v^4$$

so the inflections points are always located on the line

$$v + z = 1/2 \tag{5.11}$$

which is the asymptote when $\alpha = -1$. We notice a certain rotational sym-

metry about $z = 1/2$, $v = 0$ and in fact

$$z(v;\alpha) + z(-v;\alpha^{-1}) = 1 \tag{5.12}$$

or

$$z(v;\alpha) - 1/2 = -\{z(-v;\alpha^{-1}) - 1/2\}$$

Thus if $v-\bar{v}=\zeta z(v;\alpha)$ has a solution $v = w$ then $v-\bar{v}' = \zeta' z(v;\alpha')$ will have

the solution $v = -w$ provided

$$\alpha' = \alpha^{-1}, \ \zeta' = \zeta \ \text{ and } \ \bar{v}' = -\bar{v}-\zeta.$$

To show this we write $\alpha' = \alpha^{-1}$ and

$$v-\bar{v}' = \zeta' z(v;\alpha^{-1}) = \zeta' - \zeta' z(-v;\alpha)$$

or by replacing v by $-v$,

$$v + \bar{v}' + \zeta' = \zeta' z(v;\alpha).$$

Thus in principle we can find out everything by studying the cases for

$|\alpha| \leq 1$. It also suggests a transformation of parameters to

$$\alpha,\zeta \ \text{ and } \ w = \bar{v} + (\zeta/2) \tag{5.13}$$

118

will introduce greater symmetry since the diagrams will be invariant under $\alpha \rightarrow \alpha^{-1}$, $\zeta \rightarrow \zeta$, $w \rightarrow -w$.

We will trace the pattern of roots of eqn. (5.9) in two cases: $0<\alpha<1$ and $0>\alpha>-1$. For $\alpha>0$ there are two critical points for $\bar{v}$, namely the intersection on the v-axis of the tangents at the points of inflection, say V_+ and V_-.

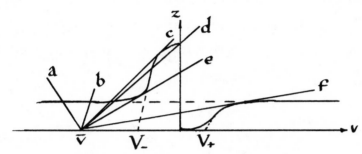

As ζ goes from $-\infty$ through 0 to ∞ the line through $\bar{v}$ swings clockwise from the horizontal through the vertical to the horizontal. Such a sequence for $\bar{v}<V_-$ is shown above as $\bar{v}a$, $\bar{v}b$,...$\bar{v}f$. Until the positive slope of $\bar{v}e$ is reached the intersection of lines such as $\bar{v}a$ and $\bar{v}b$ is unique. A line between $\bar{v}c$ and $\bar{v}d$ intersects thrice, but beyond $\bar{v}d$ only twice until $\bar{v}e$ is reached. $\bar{v}d$ is the line with $\zeta = -\bar{v}$ and, as $\bar{v} \rightarrow -\infty$, $\bar{v}c$ and $\bar{v}d$ tend to coincide. On the other hand the slope of $\bar{v}e$ is the value of z' at the point of tangency E and this tends to $\alpha/(1+\alpha)^2 y^2$ as y, the abscissa of E, tends to $-\infty$. Thus for $y \rightarrow -\infty$

$$\zeta \rightarrow (1+\alpha)^2 y^2/\alpha$$

and since $z \rightarrow \alpha/(1+\alpha)$, we also have

$$\zeta \rightarrow (y-\bar{v})(1+\alpha)/\alpha$$

and hence asymptotically we have the relation

$$\alpha\zeta = [\alpha\zeta + (1+\alpha)\bar{v}]^2.$$

When the point $\bar{v}$ is close to V_- there is only a slight difference between the slopes of $\bar{v}c$ and $\bar{v}e$. The region of two or three intersection corresponding to lines between $\bar{v}c$ and $\bar{v}e$ is traced out by computing the slope of the tangent and its intersection with the v-axis, i.e. parametically by letting v run from $-\infty$ to 0 in

$$\zeta = v^2/z(1-z), \quad \bar{v} = v-z/z' = v(1-v-z)/(1-z) \tag{5.14}$$

Similarly the region of two intersections for a line of lesser slope than $\bar{v}f$ is traced out by letting v run from 0 to ∞. If $\bar{v}$ is positive but less than V_+ there are three intersections. Thus the pattern of multiplicity in the $\zeta\,\bar{v}$ plane is as below

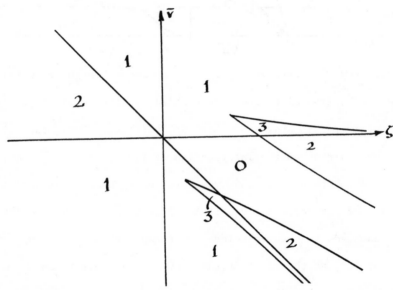

As α varies the cusps at the tip of the regions of triplicity move on the hyperbola $\bar{v}^{-2} + \bar{v}\zeta = \zeta/4$. There is a measure of symmetry about the line $\bar{v} = -\zeta$ and this confirms the previous impression that $w = \bar{v} + 1/2\zeta$ may have merit as a parameter. In the ζ, w-plane we have

120

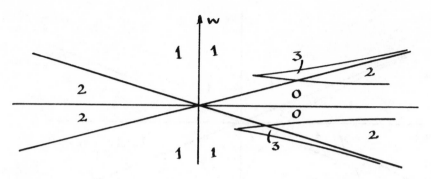

In these coordinates the locus of cusps is the hyperbola

$$4(\zeta+1/2)^2 - 16w^2 = 1.$$

The pattern for α^{-1} is obtained from that for α by rotation about the

ζ axis.

For negative α we can use a similar argument first tracing out the

ζ, $\bar{v}$-locus by eqn. (5.14). The correspondence is best shown separately.

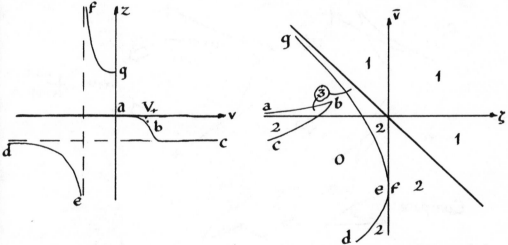

The symmetrizing effect of using ζ and w as parameters is less marked

but is shown below and the form for α^{-1} is obtained by reflection in

$w=0$.

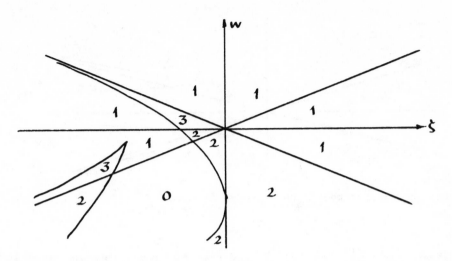

There are echoes here of the canonical catastrophes but there seems no obvious correspondence. Thus in the movement of the cusp in the case of negative α we have:

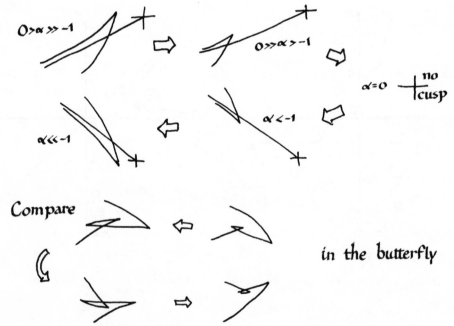

Apart from getting an enlarged view of the system of equations there may be links with similar equations in other areas. In this case, there is an analogue from statistical mechanics where Fermi-Dirac and Bose-Einstein

122

statistics show complementary changes of sign. The mean number of particles

in a state s is $\bar{n}_s$, ε_s is its energy. Then Fermi-Dirac statistics give

$$\bar{n}_s = \left\{ e^{\alpha+\beta\varepsilon_s} + 1 \right\}^{-1} = z \left(\frac{1}{\beta\varepsilon_s} ; e^{-\alpha} \right),$$

while Bose-Einstein give

$$\bar{n}_s = \left\{ e^{\alpha+\beta\varepsilon_s} - 1 \right\} = z \left(\frac{1}{\beta\varepsilon_s} ; -e^{-\alpha} \right).$$

Since β is the chemists' abbreviation for $1/kT$, $1/\beta\varepsilon_s$ is a dimensionless

temperature v.

5.3 Observable quantities.

Returning from the speculative to the practical, a word should be

inserted about the value of using observable quantities. What is meant by

this is best illustrated by an example. In Sec. 4.4 we looked at the

equations for diffusion and reaction in a catalyst pellet

$$\nabla^2 \xi + R(\xi) = 0 \quad \text{in} \quad \Omega,$$

$$\xi = 0 \quad \text{on} \quad \partial\Omega. \tag{4.66 bis}$$

To see the application of observable quantities we need a slightly different

form of the equations. Now ξ has the curious dimensions of moles per unit

length per unit time and it needs to be made dimensionless by being divided

by the product of a characteristic value of a diffusion coefficient and

concentration. Let these be D* and c* and set

$$w = \xi/D*c* \tag{5.15}$$

Also let us write $R(\xi)$ in the form

$$R(\xi) = r[c_{1s} + (\alpha_1/D_1)\xi, \ldots T_s(-\Delta H)/k_e)\xi]$$

$$= r[c_{1s} + (\alpha_1 D^*/D_1)c^*w, \ldots, T_s + \{(-\Delta H)D^*c^*/k_e\}w] \qquad (5.16)$$

$$= r(c_{1s}, \ldots T_s)P(w)$$

so that P is the ratio of the reaction rate to its value under surface conditions. Finally if the variables in the Laplacian are made dimensionless by dividing by a characteristic length L^* we have

$$\nabla^2 w + \phi^2 P(w) = 0 \text{ in } \Omega \qquad (5.17)$$

$$w = 0 \quad \text{on } \partial\Omega \qquad (5.18)$$

where

$$\phi^2 = L^{*2} \frac{r(c_{1s}, \ldots T_s)}{D^*c^*} \qquad (5.19)$$

This last parameter (often known as the Thiele modulus in honour of one of the pioneers in this field) is a measure of the ratio of the reaction rate to the diffusion rate. There will of course be other parameters in $P(w)$ but, supposing these to be constant, the solution of eqn. (5.17) will be a function of ϕ.

Now there is a very important functional of the solution, namely

$$\eta = \frac{1}{V_\Omega} \iiint_\Omega P(w)dV. \qquad (5.20)$$

the average value of the reaction rate in the pellet as a ratio of the surface reaction rate. This so-called "effectiveness factor" is a function of the parameter ϕ and wraps up the practical implications of the solution very neatly. As $\phi \to 0$, $\eta \to 1$, for eqn. (5.17) becomes Laplace's equation and, with the boundary condition (5.18), the solution is $w=0$; but $P(0)=1$ so $\eta=1$. Physically this makes sense, for $\phi \to 0$ means that diffusion is rapid in comparison with reaction and so the surface conditions prevail everywhere. As $\phi \to \infty$, $\eta \to 0$ since, with increasingly rapid reaction and slower diffusion,

124

the interior of the pellet is starved of reactants and ineffective. In fact a singular perturbation analysis shows that η is inversely proportional to ϕ for large ϕ. The log-log plot of $\eta(\phi)$ has the form shown in the simplest

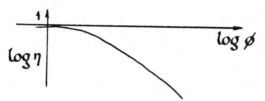

cases. For example if $P(w)=1-w$, $\eta(\phi)=(\tanh\phi)/\phi$ when Ω is a flat plate of thickness $2L*$ with sealed edges; it is $\eta(\phi) = (3/\phi^2)(\phi\coth\phi-1)$ for a sphere of radius $L*$.

Now it is all very well to know $\eta=\eta(\phi)$ but when the reaction rate is measured it is not $r(c_{1s},\ldots T_s) = r_s$ that is observed, but rather $\eta r_s = r_{obs}$ since the intrusive effects of diffusion are present in the measuring process. It is therefore useful to define an observable parameter

$$\phi^2 = L*^2 r_{obs}/D*c* = \eta\phi^2 \qquad (5.21)$$

and to plot η as a function of Φ rather than of ϕ. We notice that Φ and ϕ are virtually the same when they are small but that Φ is proportional to the square root of ϕ, when they are large. Thus the graph of $\eta(\Phi)$ is a little steeper and so defines a little better than $\eta(\phi)$ a value

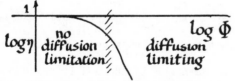

of Φ above which diffusion limitation can be regarded as serious. There is of course something arbitrary about such a critical value (it might for instance, be the value at which $\eta = 0.8$, or $\eta = 0.9$ if one wanted to be more cautious), but it can be a very sensible arbitrariness and once determined can be correlated with other parameters. Bischoff [30] has shown how to normalize ϕ so that a sensible critical value is $\phi=1$.

The notion of a normalized modulus is an elegant one and often worth cultivating. For example, we have noted that the asymptotic form of η is inversely proportional to ϕ. The constant of proportionality depends on the shape of Ω and on the kinetics. However if w is scaled so that $P(1)=0$ and $L*$ is chosen so that

$$L* + \frac{V}{S} \left\{ \int_0^1 P(w)\,dw \right\}^{1/2}, \tag{5.22}$$

where V is the volume and S the surface area of Ω, then the constant of proportionality is 1 (i.e. $\eta \sim \phi^{-1}$) for all shapes and kinetics. The basis for this statement lies in the singular perturbation solution of eqn. 5.17 and will not be pursued here, but it can readily be appreciated that when ϕ is large only a thin layer beneath the surface is contributing to the reaction and hence the surface area if important. Details can be found in [14].

5.4 Comparison of models and prototypes and of models among themselves.

When a model is being used as a simulation an obvious comparison can be made between its predictions and the results of experiment. We are favourably impressed with the model if the agreement is good and if it has not been purchased at the price of too many empirical constants adjusted to fit the data. If the parameters are determined independently and fed into the final model as fixed constants not to be further adjusted, then we can have a fair degree of confidence in the data and in the model. Both model and data have their own integrity the former in the relevance and clarity of its hypotheses and the rigour and appropriateness of its development, the latter in the carefulness of the experimenter and the accuracy of the results. But these virtues do not only inhere in the possessors they also gain validity from the other. Thus, as Truesdell remarks, in applied mathematics rigour is of the essence, for the comparison can have no meaning if the model has not been handled properly. Similarly data must be of a certain degree of accuracy or it has no ability to prove (i.e. test) a theoretical viewpoint. Thus the attitude of never believing an experiment until its confirmed by theory has as much to be said for it as that which never believes a theory before its confirmation by experiment.

In the comparison of theory with experiment an array of statistical tools is available and should be used. Thus not only can the fitted constants be chosen in some best sense (e.g. least squares) but it is not difficult to find also the covariance matrix of the estimates and hence detect any hidden sensitivities. One danger that is easy to overlook is the existence of hidden constancies that will give spurious values. Thus the temperature rise in an adiabatic bed is a measure of the reaction rate which will be a function of the mean temperature. But if the inlet temperature is virtually constant, the observed pairs of temperature rise and mean temperature will be perfectly correlated by a straight line whatever the functional relationship between them. This straight line says no more than that the mean temperature equals the inlet plus half the rise. The classic correlation between the intelligence of the children and the drunkenness of the parents which so confounded temperance societies years ago--until it was discovered that all the data came from schools in the east end of London--is another illus- tration of a data base too narrow to test a model.

In discriminating between models it is not entirely satisfactory to fit the constants of each and choose the better fitting model. For one thing there may be little to choose between the goodness of fit in the two cases. We are on much firmer ground if the two models can be presented in such a way that they have qualitatively different behaviour. Tanner [171] has

127

tried plotting data on the intermediates of a complex reaction in such a way that they fall on a loop which in some models is traversed clockwise and in others anti-clockwise.

A field in which qualitative behaviour of models has been used discriminatingly is the study of oscillations in chemical reaction. If a certain reaction is known to give oscillations under certain conditions then any mechanism that is incapable of giving oscillations under these conditions is ruled out. Sheintuch and Schmitz have reviewed this subject very thoroughly and examined the models for the oxidation of carbon monoxide in the light of this criterion, [163]; see also Eigenberger [61] and Luss and Pikios [124]. They find that a more than usually detailed account of the mechanism is needed. Unless the lack of uniformity of the catalyst surface, the affects of the chemisorbed species, their variations of reactivity and the dependence of the activation energy on coverage are brought into consideration no oscillations can occur. This rules out seven of the thirteen cases they consider and some of the remaining possibilities are seen to be unlikely by the magnitude that various terms would have to have if oscillations were to occur. This is remarkable in that twelve of the thirteen cases can be adjusted to match the known form of the reaction rate expression for carbon monoxide oxidation.

This work of Schmitz and Sheintuch shows the power of mathematical modelling when combined with physical understanding, for not only are many possibilities eliminated, but the features that call for further investigation in the remaining candidate models are clearly brought out. The mention of it is a pleasant note on which to conclude this exploration of the craft of mathematical modelling.

Appendix A Longitudinal diffusion in a packed bed

Description of the system P.

A fluid flows through the interstices of a long cylinder which is packed with particles. Because of the variations in local velocity as the fluid passes around the particles, the eddying and wall effects, some molecules will pass through more quickly than others. In addition there is the molecular diffusion of the tracer molecules in the flow field. Hence, if the fluid is marked with a tracer which enters the bed at time $t = 0$ in a sharp pulse and the concentration of the tracer is measured as the stream leaves the bed,

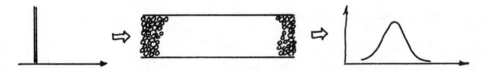

it will be found to have spread out into a diffuse band. If the input were perfectly instantaneous and the output $C(t)$ then

$$\mu = \int_0^\infty tC(t)dt / \int_0^\infty C(t)dt$$

would be the mean residence time of tracer molecules and

$$\sigma^2 = \int_0^\infty (t-\mu)^2 C(t)dt / \int_0^\infty C(t)dt$$

the variance of residence times. Clearly these two statistics of $C(t)$ give some idea of the dispersive effect of the interaction of these physical processes though the equations should be capable of yielding $C(t)$ itself. We denote by Ω the intersticial space through which the fluid flows, by $\partial\Omega$ the bounding surface of this region excluding $\partial\Omega_i$ and $\partial\Omega_o$, the inlet and outlet ends of the cylinder. Let the area of cross-section of the

129

cylinder (and thus of $\partial\Omega_i$ and $\partial\Omega_o$) be A, q the volumetric flow rate, L the length and ε the fractional free space. The mean linear velocity of flow is thus $q/A\varepsilon = L/\Theta$, where Θ is defined by this equation as a residence time. The concentration of tracer at any point $\underset{\sim}{r}$ of the free space and time t is $c(\underset{\sim}{r},t)$; $C(t) = \iint_{\partial\Omega_o} c(r,t)dS$.

Hypotheses.

Physical laws and general conservation principles will be regarded as an underlying hypothesis, H_o. Then the specific hypotheses we need are as follows:

H_1: the fluid is incompressible and the flow steady.

H_2: its motion is governed by the Navier-Stokes equations.

H_3: the diffusion of the tracer obeys Fick's law.

H_4: the tracer does not penetrate the cylinder walls or enter the particles.

H_5: the mean linear velocity is uniform.

H_6: the dispersive effect is given by an effective longitudinal diffusion coefficient.

H_7: the movements of a tracer particle can be thought of in discrete time intervals during which it either moves forward by an increment of length or, being caught in an eddy, moves not at all.

Model Π_1.

This model is just the embodiment of the hypotheses H_o-H_4 using well-known equations; we shall not go into the derivation of the Navier-Stokes and similar equations but use them as needed; they are of course themselves derived by the principles outlined in Ch. 3. Thus, if $\underset{\sim}{v}(\underset{\sim}{r},t)$ is the

130

velocity at some point $\underset{\sim}{r}$ in Ω, H_1 implies that it is really $\underset{\sim}{v}(\underset{\sim}{r})$ and the incompressibility that

$$\nabla \cdot \underset{\sim}{v} = 0 \tag{A1}$$

The Navier-Stokes equations are

$$(\underset{\sim}{v} \cdot \nabla)\underset{\sim}{v} = -\nabla\bar{p}/\rho + \nu\nabla^2\underset{\sim}{v}. \tag{A2}$$

and these are also written for a steady flow with $\bar{p}$ the pressure, ρ the density and ν the kinematic viscosity. The equation governing the dispersion of the tracer is not a steady equation but, with D the molecular diffusion coefficient, is

$$\frac{\partial c}{\partial t} + \underset{\sim}{v} \cdot \nabla c = D\nabla^2 c. \tag{A3}$$

These three equations are subject to the boundary conditions

$$\underset{\sim}{v} = 0 \text{ on } \partial\Omega, \ \bar{p} = \bar{p}_i, \ \underset{\sim}{v} = \underset{\sim}{v}_i \text{ on } \partial\Omega_i \tag{A4}$$

and if $\underset{\sim}{n}$ denotes the outward normal to the boundary of Ω,

$$\underset{\sim}{n} \cdot \nabla c = 0 \text{ on } \partial\Omega \text{ or } \partial\Omega_o,$$

$$\tag{A5}$$

$$\underset{\sim}{v}c - D\nabla c = \underset{\sim}{v}c_i \text{ on } \partial\Omega_i.$$

If the impulse of tracer put to the bed is perfectly sharp we might put

$$c_i = \delta(t)/q \tag{A6}$$

since then $\int_0^\infty \iint_{\partial\Omega} c_i \underset{\sim}{v} \cdot \underset{\sim}{n} \ dS \ dt = 1$. Initially the bed is free of tracer so

$$c(\underset{\sim}{r}, o) = 0 \tag{A7}$$

Model II_2.
_____

The first model is so complicated as to be almost impossible of solution and clearly the complexity of the geometry is part of the problem. Suppose we take the drastic step of ignoring this complexity and say that the flow averages out to a virtually uniform velocity $U = q/A\varepsilon$ and that the

131

dispersion is the sort of thing we would see if Fickian diffusion were imposed on this. The prototype has thus been modified by the hypotheses, H_5, H_6, to an equivalent continuum. The equations for this are for a concentration $c(x,t)$ which is a function of x the distance from the inlet and time. Then

$$\frac{\partial c}{\partial t} + U \frac{\partial c}{\partial x} = D_e \frac{\partial^2 c}{\partial x^2}, \; 0 \le x \le L, \tag{A8}$$

where D_e is the effective diffusion coefficient. As boundary conditions we have

$$Uc = D_e \frac{\partial c}{\partial x} = Uc_i, \; x = 0 \tag{A9}$$

and

$$\frac{\partial c}{\partial n} = 0, \; x = L. \tag{A10}$$

Initially,

$$c(x,o) = 0. \tag{A11}$$

The measured output is

$$C(t) = c(L,t). \tag{A12}$$

Model II_3.

In formulating II_3 we take the opposite view of the physical system from the continuous analogue of P_2 and emphasize the discreteness of the packed bed. In particular the flow is a seaweed flow, squeezing between particles and bulging into the cavities. For example, in a rhombohedral blocked passage arrangement the overall fractional free volume is 26% but the free area perpendicular to the flow varies widely. Through a plane of centres of the spheres it is only 9% while if the region between one plane of centres and the next is divided into thirds the average free area is 41% in the middle third but only 18% in the other two. It is not entirely unreasonable

132

therefore to regard each layer of particles as a cell in whose interstices, a thorough mixing takes place. The system P is thus replaced by P_3, a sequence of N cells each of volume V through which the stream passes. Equating the total free volume gives $NV = AL\varepsilon$. We will not be dogmatic about N at this point, though, if the analogy holds up we might expect N to be of the order of L/d_p where d_p is the particle diameter.

If $c_n(t)$ is the concentration of tracer in cell n

$$V \frac{dc_n}{dt} = q(c_{n-1} - c_n) \tag{A13}$$

with

$$V \frac{dc_1}{dt} = q(c_i - c_1) \tag{A14}$$

and

$$c_n(0) = 0 \tag{A15}$$

The observed quantity

$$C(t) = c_n(t). \tag{A16}$$

This model has been elaborated to consider radial as well as longitudinal dispersion in packed beds by Deans and Lapidus [53].

The connection between Π_2 and Π_3.

As is remarked in Ch. 1, there is no immediate connection between Π_2 and Π_3 for the discretization of Π_2 would not produce Π_3. They are connected only in the sense that they give comparable solutions. Thus if $c_i(t)$ is given the solution of Π_2 is

$$C(t) = \int_0^t c_i(t') p_2(L, t-t') dt' \tag{A17}$$

where

$$p_2(x,t) = U(\pi D_e t)^{-1/2} \exp- \frac{(x-Ut)^2}{4D_e t} - \frac{U^2}{2D_e} \exp \frac{Ux}{D_e} \operatorname{erfc} \frac{x+Ut}{2(D_e t)^{1/2}} \qquad (A18)$$

In particular if $c_i(t)$ is the delta function input, the output is $p_2(L,t)$ and $\int_0^t p_2 dt = 1$. The distribution of residence times is almost Gaussian and the mean and variance of residence times are

$$\mu_2 = \frac{L}{U}\left[1 + \frac{1}{P}\right] \qquad (A19)$$

$$\sigma_2^2 = \frac{2DL}{U^3}\left[1 + \frac{3}{4P}\right] = \frac{L^2}{U^2 P}\left[1 + \frac{3}{4P}\right] \qquad (A20)$$

where $P = UL/2D_e$.

The solution for Π_3 is

$$C(t) = \int_0^t c_i(t') p_3(t-t') dt' \qquad (A21)$$

where

$$p_3(t) = \frac{1}{(N-1)!}\left[\frac{q}{v}\right]^N t^{N-1} e^{-qt/v} \qquad (A22)$$

This is the Poisson distribution which is also asymptotically Gaussian and whose mean and variance are

$$\mu_3 = NV/q \qquad (A23)$$

$$\sigma_3^2 = NV^2/q^2 \qquad (A24)$$

Since the total volume of the cells, NV, should obviously equal the total free space, $LA\varepsilon$, and $q = A\varepsilon U$, we see that

$$\mu_3 = NV/q = L/U = \Theta$$

which is approximately equal to μ_2 if $P \gg 1$. But if we equate μ^2/σ^2 in the two cases we see that

$$N = P(1 + \frac{1}{P})^2 / (1 + \frac{3}{4P})$$

$$\doteq P + \frac{5}{4}.$$

As we have seen, there are physical grounds for thinking that N should be of the order of L/d_p which is quite large; hence $P \doteq N \gg 1$ and the means and variances of the two models are approximately the same. If $P = N = L/d_p$ then the so-called particle Peclet number $Pe = Ud_p/D_e = 2$ which is in good agreement with experiment. An alternative way of establishing the connection between Π_2 and Π_3 by the common Gaussian approximation to P_2 and P_3 was discussed by Amundson and Aris [5].

Model Π_4.

A rather different model is obtained by taking a disjointed view of the tracer movement and saying that sometimes it moves forward with the stream and at others it is caught in an eddy and stays in virtually the same place. This is the crudest of random walk assumptions, embodied in H_7, and obviously could be elaborated by giving a distribution of lengths over which the movement might take place. Suppose that in each interval of time τ the particle either moves forward a distance δ or remains where it is. The probability of the first event is p and of the second is $q = 1-p$. During a time $t = M\tau$ there have been M such "choices" and, if $L = N\delta$, the particle will emerge if N of these have been to move forward. The probability of this is

$$P_4 = \begin{bmatrix} M \\ N \end{bmatrix} p^N q^{M-N} \tag{A25}$$

Thus

$$C(t) = \begin{bmatrix} t/\tau \\ L/\delta \end{bmatrix} p^{(L/\delta)} q^{(t/\tau)-(L/\delta)} \tag{A26}$$

It is well known that if

$$z = \frac{N-Mp}{(Mpq)^{1/2}} = \frac{L-(\delta/\tau)tp}{\delta(tpq/\tau)^{1/2}} \tag{A27}$$

then

$$C(\tau) \sim \frac{1}{(2\pi tpq/\tau)^{1/2}} \exp - \frac{(L-(\delta p/\tau)t)^2}{2(\delta^2 pq/\tau)t} \qquad (A28)$$

as N and $M \to \infty$ and $z^3 M^{-1/2} \to 0$. Comparing this with (A18) we see that δ, τ and p should satisfy

$$\delta p/\tau = U, \quad \delta^2 pq/\tau = D_e. \qquad (A29)$$

Thus if δ is of the order of d_p and $Pe = Ud_p/D_e = 2$ we have $q = p = 1/2$, which is reasonable enough.

Notation for the models II.

A	cross-sectional area of bed
$C(t)$	outcoming average concentration
$c(\underset{\sim}{r},t)$	concentration at position $\underset{\sim}{r}$ and time t
$c_i(t)$	input concentration
$c_n(t)$	concentration in n^{th} cell
D	molecular diffusion coefficient of tracer
D_e	effective or equivalent diffusion coefficient in II_2
d_p	particle diameter
L	length of packed bed
M	number of time increments in II_4
N	number of cells in II_3 or space increments in II_4
P	$UL/2D_e$
Pe	Ud_p/D_e
p	probability of movement in II_4, $q = 1-p$
P_2, P_3	residence time probability densities in II_2 and II_3
$\bar{p}, \bar{p}_i$	pressure, inlet pressure in II_1
q	volumetric flow rate
$\underset{\sim}{r}$	reactor of position in free space of packed bed

136

t	time
U	linear velocity
V	volume of cell
$\underset{\sim}{v}$	reactor of velocity in interstices of bed
x	distance from inlet
δ	length of step in Π_4
ε	fractional free volume of bed
Θ	residence time
μ_i	mean residence time in Π_i, $i = 2,3$
ν	kinematic viscosity
ρ	density
σ_i^2	variance of residence times in Π_i, $i = 2,3$
τ	time increment in Π_4
Ω	free space of packed bed
$\partial\Omega$	boundary of Ω except for $\partial\Omega_i$ and $\partial\Omega_o$
$\partial\Omega_i$, $\partial\Omega_o$	inlet and outlet boundaries

Appendix B The coated tube chromatograph and Taylor diffusion

Description of the systems C and D.

The inside of a long, cylindrical tube is coated with a thin retentive layer. A carrier gas flows through the tube and molecules of a tracer solute are convected by it, they diffuse and, if they reach the wall, may pass into the retentive layer and spend some time there. Different tracer solutes with different affinities for the retentive layer will spend different proportions of their time in this stationary phase and so peaks of different solutes are separated emerging at different times. (The analogy is sometimes made with a stream of soldiers on a long road lined with pubs: the teetotalers will arrive at the end of the road first, followed by the temperate and the topers--the dipsomaniacs may never make it.) But the diffusion, the variations of velocity across the tube and the rate of partition between the phases will all contribute to the spreading out of an initially sharp peak. Our interest is to account for the mean speed of the peak and to understand how each factor affects its spread.

There is also the special subordinate case when there is no retentive layer (system D). Here attention is focussed entirely on the interaction of diffusion and convection. This is the so-called Taylor diffusion problem first successfully analyzed by Sir Geoffrey Taylor [174]. Lateral diffusion prevents the solute from travelling with any one streamline and counteracts the spreading effect of the wide variation of flow rates. Thus there is a Taylor diffusion coefficient which is inversely proportional to the molecular diffusion coefficient. An immense literature on this problem now

138

exists which is it not our purpose to summarize here: an early summary was given by Taylor [175] and a later one by Gill and Nunge [75] but there has not been any survey of the most recent developments. Extensions to the chemical reactor where reaction takes place in the fluid, in the retentive layer or in both have also been proposed.

Hypotheses of systems C and D.

Again H_o will be taken to embrace the underlying scientific laws and the following are the specific assumptions that are introduced to define various models.

H_1: the tube in infinite in both directions

H_2: the tube stretches from the origin, x=0, to infinity

H_3: the tube is finite, $0 \le x \le L$.

H_4: the velocities in the cylinder, $0=r_o \le r \le r_1$, and the annulus, $r_1 \le r \le r_2$, are functions only of r, the distance from the axis say $U_i \phi_i(r), i=1,2$, where U_i is the mean velocity in the region.

H_5: the annular region (retentive coating) is stationary.

H_6: the coating is thin, $(r_2-r_1) << r_1$.

H_7: there is no coating.

H_8: the rate of exchange between the two regions is proportional to the difference $(c_2-\alpha c_1)$, where c_1 and c_2 are the concentrations in the cylinder and annulus and α a constant.

H_9: the diffusion coefficients D_i, i = 1,2, are constant (a more general hypothesis is made in Aris [11] but we need not reach for too great a generality).

H_{10}: there is no flow across the axis, $r = r_o=0$, or the outer wall, $r=r_2$.

H_{11}: sufficient insight into the problem is to be gained from the temporal evolution of the moments of the distribution in space.

H_{12}: sufficient insight is to be had from calculating the mean concentration across the tube and determining an effective speed and dispersion coefficient.

H_{13}: longitudinal diffusion is unimportant.

H_{14}: the flow profile is parabolic $\phi_1(r) = 2[1-(r/r_1)^2]$.

H_{15}: diffusion plays no role.

The most general model Γ_1.

Clearly some of the foregoing hypotheses are mutually contradictory and will be used to define different case and sets of boundary conditions. The basic equations of convection and diffusion (i.e. H_0) with H_4 and H_9 give

$$\frac{\partial c_i}{\partial t} + U_i \phi_i(r) \frac{\partial c_i}{\partial x} = D_i \left\{ \frac{\partial^2 c_i}{\partial x^2} + \frac{1}{r} \frac{\partial}{\partial r} \left[r \frac{\partial c_i}{\partial r} \right] \right\} , \quad i = 1,2 \tag{B1}$$

The hypothesis H_{10} gives us the boundary conditions

$$D_1 \frac{\partial c_1}{\partial r} = 0, \ r = 0, \tag{B2}$$

and

$$D_2 \frac{\partial c_2}{\partial r} = 0, \ r = r_2, \tag{B3}$$

while H_8 gives

$$D_1 \frac{\partial c_1}{\partial r} = D_2 \frac{\partial c_2}{\partial r} = k(c_2 - \alpha c_1), \ r = r_1. \tag{B4}$$

The initial conditions have to be specified and are

$$c_i(x,r,o) = \chi_i(x,r), \ i = 1,2. \tag{B5}$$

140

We have not said anything yet as to the boundary conditions with respect to x. If H_1 is asserted we need only add

$$c_i(x,r,t) \text{ is finite as } x \to \pm \infty \qquad\qquad (B6)$$

and the problem is complete. Γ_1 is the set of equations (B1) - (B6).

It proves advantageous in the analysis of this problem to have a moving origin, so that alternative equations to (B1) are

$$y = x-Vt \qquad\qquad (B7)$$

$$\frac{\partial c_i}{\partial t} + (U_i \phi_i(r)-V)\frac{\partial c_i}{\partial y} = D_i \left\{ \frac{\partial^2 c_i}{\partial y^2} + \frac{1}{r}\frac{\partial}{\partial r}\left(r\frac{\partial c_i}{\partial t}\right) \right\}. \qquad (B8)$$

This is not a new model but a preliminary modification, Γ_1'.

The models Γ_2 and Γ_3.

The doubly infinite tube is not a very accurate model of the practical situation. It can be realized by a long capillary tube with the variation of tracer concentration confined to a comparatively narrow range far from the ends. Indeed experiments have been done by flashing light on a short section of a tube through which the flow of a light sensitive fluid passes. Such a flash at t=0 established the initial concentration distribution χ and the conditions of Γ_1 apply very precisely. A far more realistic situation however is afforded by replacing H_1 by H_2. It then becomes necessary to replace the condition of finiteness as $x \to -\infty$ by an inlet condition at x = 0. If U_1 and U_2 are both positive and if it is felt that a solute fed to the plane x = 0 (for example from a reservoir) must enter the system then

$$D_i \frac{\partial c_i}{\partial x} + U_i \phi_i(r)\{c_i - c_{if}\} = 0, \; i=1,2 \qquad\qquad (B9)$$

at x = 0, where c_{if} is the feed concentration.

If $U_2 < 0$, as when a thin film runs down the wall of a vertical cylinder countercurrent to an upward flow of gas then we have to invoke H_3 since the countercurrent film must be specified at its inlet and write

$$D_i \frac{\partial c_1}{\partial x} + U_1 \phi_1(r) \{c_1 - c_{1f}\} = 0, \quad x=0, \quad U_1 > 0,$$

$$D_2 \frac{\partial c_2}{\partial x} + U_2 \phi_2(r) \{c_2 - c_{2f}\} = 0, \quad x=L, \quad U_2 < 0 \qquad \text{(B10)}$$

With the semifinite tube there is still the requirement of finiteness as $x \to \infty$, but for the finite tube another condition is required. These are often written

$$\frac{\partial c_1}{\partial x} = 0, \quad x=L; \quad \frac{\partial c_2}{\partial x} = 0, \quad x=0. \qquad \text{(B11)}$$

The moment model Γ_4.
_____

The important characteristic of all dispersion situations is that there is a certain movement of the centre of gravity of the solute distribution and a steady increase in its spread. It follows that some insight may be gained by studying the temporal evolution of the spatial moments. If we had knowledge of <u>all</u> the moments then, with suitable restrictions, we should have knowledge of the whole distribution (see e.g. [162]), but, apart from the unlikely event of there being a special form of solution, this demands solution of an infinite number of equations and so is a worse task than the solution of Γ_1 itself. Of course, our hope is that the first two or three moments will tell us all we want to know.

Let

$$c_i^{(p)}(r,t) = \int_{-\infty}^{\infty} y^p c(y,r,t)\,dy \qquad \text{(B12)}$$

be the pth moment about the origin moving with speed V and

$$m^{(p)}(t) = \int_0^{r_1} 2rc_1^{(p)}(r,t) + \int_{r_1}^{r_2} 2rc_2^{(p)}(r,t)dr. \tag{B13}$$

Then

$$\frac{\partial c_i^{(p)}}{\partial t} = D_i \frac{1}{r} \frac{\partial}{\partial r}\left[r \frac{\partial c_i^{(p)}}{\partial r}\right] + D_i p(p-1)c_i^{(p-2)}$$

$$+ \{U_i \phi_i(r)-V\}pc_i^{(p-1)} \tag{B14}$$

with

$$\frac{\partial c_i^{(p)}}{\partial r} = 0, \ r=0; \quad \frac{\partial c_2^{(p)}}{\partial r} = 0, \ r=r_2 \tag{B15}$$

and

$$D_i \frac{\partial c_1^{(p)}}{\partial r} = D_2 \frac{\partial c_2^{(p)}}{\partial r} = k(c_2^{(p)}-\alpha c_1^{(p)}), \ r=r_1 \tag{B16}$$

Initially

$$c_i^{(p)}(r,0) = \chi_i^{(p)}(r) = \int_{-\infty}^{\infty} x^p \chi_i(x,r)dx \tag{B17}$$

and it is of course assumed that these moments are finite.

These equations come easily from multiplying the earlier equations by y^p and integrating from $-\infty$ to ∞. To get equations for $m^{(p)}$ we have to average over the cross-section, giving

$$\frac{dm^{(p)}}{dt} = p(p-1)\sum_1^2 D_i \int_{r_{i-1}}^{r_i} 2rc_i^{(p-2)}(r)dr + p \sum_1^2 \int_{r_{i-1}}^{r_i} 2r\{U_i \phi_i(r)-V\} c_i^{(p-1)}dr \tag{B18}$$

with

$$m^{(p)}(0) = \sum_1^2 \int_{r_{i-1}}^{r_i} 2r\chi_i^{(p)}(r)dr. \tag{B19}$$

It is perhaps well to interpolate here the results for Γ_4 since they can be misunderstood. Without loss of generality we can take the initial distribution to be such that $m^{(0)}(0)=1$, $m^{(1)}(0)=0$. It can then be shown

143

[11] that $m^{(0)}(t)=1$, $c_1^{(0)}(t)\to\pi/(A_1+\alpha A_2)$, $c_2^{(0)}(t)\to\alpha\pi/(A_1+\alpha A_2)$ where $A_i=\pi(r_i^2-r_{i-1}^2)$. Moreover if V is chosen to be $(A_1U_1+\alpha A_2U_2)/(A_1+\alpha A_2)$ then $dm^{(1)}/dt\to 0$. This means that the centre of gravity of the solute ultimately moves with the weighted mean speed of the two streams, the weights being the amounts of the solute in the two phases. This is very reasonable but it should be emphasized that it is an asymptotic result and, while one can give estimates of how quickly it is approached, the exact value of $m^{(1)}$ depends on the initial distribution. If $\beta_1=A_1/(A_1+\alpha A_2)$, $\beta_2=1-\beta_1$, the asymptotic result for $m^{(2)}$ is that its growth rate

$$\frac{dm^{(2)}}{dt} \to 2\sum_1^2\beta_i\left[D_i + \kappa_i\left\{\frac{U_i^2(r_i^2-r_{i-1}^2)}{D_i}\right\} + \frac{\beta_1^2\beta_2^2(A_1+\alpha A_2)(U_1-U_2)^2}{\pi k \alpha r_1}\right] \qquad (B20)$$

where

$$\kappa_i = \kappa_{i1}-2\kappa_{i2}(V/U_i)+\kappa_{i3}(V/U_i)^2 \qquad (B21)$$

and the κ_{ij} can be calculated from a knowledge of the velocity profiles. For example, when $U_2=0$ (hypothesis H_5) and $r_2\to r_1$ (H_6) the first two terms are $\beta_1[D_1+(U_1^2r_1^2/48D_1)(11-16\beta_1+6\beta_1^2)]$ and $\beta_2[D_2+(U_1^2(r_2-r_1)^2/3D_2)(1-2\beta_2+\beta_2^2)]$ respectively. Again these growth rates are only asymptotically valid and nothing more is rigorously claimed for this approach. It is quite another step to draw comparisons with apparent longitudinal dispersion coefficients such as is done in subsequent models.

The Taylor diffusion models with laminar flow.

We now specialize from the chromatographic case to the case to Taylor diffusion in laminar flow, i.e. we invoke H_7 and H_{14}. As before, we could let the system be much more general (cf. [10]) but this is not to the point here. We then have Δ_1 corresponding to the most complete equations

144

for the physical system D. We can drop the suffixes since there is only one region and concentration and will write $r_1 = a$. Then

$$\frac{\partial c}{\partial t} + 2U \left[1 - \frac{r^2}{a^2}\right] \frac{\partial c}{\partial x} = D \frac{1}{r} \frac{\partial}{\partial r} \left[r \frac{\partial c}{\partial r}\right] + D \frac{\partial^2 c}{\partial x^2} \tag{B22}$$

and

$$\frac{\partial c}{\partial r} = 0, \ r = 0, a \tag{B23}$$

$$c(x,r,t) \text{ finite as } x \to \pm\infty \tag{B24}$$

$$c(x,r,0) = \chi(x,r) \tag{B25}$$

This is the system Δ_1 which can be modified to Δ_1' by the change of variable

$$y = x - Ut \tag{B26}$$

giving

$$\frac{\partial c}{\partial t} + U \left[1 - 2 \frac{r^2}{a^2}\right] \frac{\partial c}{\partial y} = D \frac{1}{r} \frac{\partial}{\partial r} \left[r \frac{\partial c}{\partial r}\right] + D \frac{\partial^2 c}{\partial y^2} . \tag{B27}$$

Gill et al. [74] invoked H_2 and assumed that $c(o,r,t)$ could be specified, in fact as a constant which without loss of generality can be taken to be 1. Thus to (B22) and (B23) there is added

$$c(o,r,t) = 1, \tag{B28}$$

$$c(x,r,t) \to 0 \text{ as } x \to \infty, \tag{B29}$$

$$c(x,r,0) = 0 \tag{B30}$$

This is Δ_2 and is worth noting that Gill made them nondimensional in a way that would bring out the scale of the spread into which the sharp front at $t=0$, $x=0$ will soften as t increases. Thus with

$$\rho = \frac{r}{a}, \xi = \frac{x}{2U} \sqrt{\frac{D}{t}}, \ \tau = \frac{Dt}{a^2}, \ P = \frac{2aU}{D} \tag{B31}$$

we have

$$\frac{\partial c}{\partial \tau} + \left[\frac{1-\rho^2}{\frac{1/2}{\tau}} - \frac{\xi}{2\tau}\right]\frac{\partial c}{\partial \xi} = \frac{\partial^2 c}{\partial \rho^2} + \frac{1}{\rho}\frac{\partial c}{\partial \rho} + \frac{1}{4\tau P^2}\frac{\partial^2 c}{\partial \xi^2} \qquad (B32)$$

with

$$\frac{\partial c}{\partial \rho} = 0, \ \rho = 0,1, \qquad (B33)$$

$$c(0,\rho,\tau) = 1, \qquad (B34)$$

$$c(\xi,\rho,\tau) \to 0 \text{ as } \xi \to \infty. \qquad (B35)$$

If H_{11} is invoked and we define

$$c^{(p)}(r,t) = \int_{\infty}^{\infty} y^P c(y,r,t)\,dy$$

$$m^{(p)}(t) = \frac{1}{a^2}\int_0^a 2r c^{(p)}(r,t)\,dr \qquad (B36)$$

then

$$\frac{\partial c^{(p)}}{\partial t} = D\frac{1}{r}\frac{\partial}{\partial r}\left[r\frac{\partial c^{(p)}}{\partial r}\right] + Dp(p-1)c^{(p-2)} + pU\left[1 - \frac{r^2}{a^2}\right]c^{(p-1)} \qquad (B37)$$

and

$$\frac{dm^{(p)}}{dt} = Dp(p-1)m^{(p-2)} + p\frac{U}{a^2}\int_0^a 2r(1-\frac{r^2}{a^2})c^{(p-1)}(r,t)\,dr. \qquad (B38)$$

If $m^{(0)}(0) = 1$ and the $m^{(p)}(0)$ are finite then $m^{(0)}(t) = 1$, $m^{(1)}(t) \to a$ constant and $dm^{(2)}(t)/dt \to 2[D+(a^2 U^2/48D)]$. These are again asymptotic results for this model, Δ_3. The detailed asymptotic approach to normality of the distribution of solute has been discussed in an excellent paper by Chatwin [42].

Models with the mean concentration.

We now have the bolder hypothesis that the situation can be sufficiently represented in terms of plug flow and an effective longitudinal diffusion coefficient which govern the mean concentration

146

$$\bar{c}(x,t) = \frac{1}{a^2} \int_0^a 2rc(x,r,t)dr. \tag{B39}$$

Using Gill's boundary conditions we write

$$\frac{\partial \bar{c}}{\partial t} + U \frac{\partial \bar{c}}{\partial x} = D_e \frac{\partial^2 \bar{c}}{\partial x^2} \tag{B40}$$

with

$$\bar{c}(0,t) = 1 \tag{B41}$$

$$\bar{c}(x,t) \to 0 \text{ as } x \to \infty \tag{B42}$$

$$\bar{c}(x,0) = 0 \tag{B43}$$

This is Δ_4 and we observe that is is cognate with Δ_1 but not derivable from it, for averaging over the cross-section gives

$$\frac{\partial \bar{c}}{\partial t} + U \frac{\partial <c>}{\partial x} = D \frac{\partial^2 \bar{c}}{\partial x^2}$$

where

$$<c> = \frac{4}{a^2} \int_0^a r \left[1 - \frac{r^2}{a^2} \right] c(x,r,t)dr \tag{B44}$$

is not $\bar{c}$, but the cup-mixing mean. By subtraction

$$D_e = D + U \frac{\partial}{\partial x} [\bar{c} - <c>] \Big/ \frac{\partial^2 \bar{c}}{\partial x^2} \tag{B45}$$

and this equation has been used to calculate a variable D_e from any particular solution.

The model Δ_4 may be varied by the choice of D_e and we recognize the following possibilities. If H_{13} is invoked then Taylor [174,175] showed that

$$\Delta_4' : D_e = \frac{a^2 U^2}{48D} \tag{B46}$$

The growth of moments in Δ_4 and Δ_3 are asymptotically the same if

147

$$\Delta_4'': \quad D_e = D + \frac{a^2 U^2}{48D} \tag{B47}$$

On the other hand diffusion may dominate completely in a very slow flow and

$$\Delta_4''': \quad D_e = D. \tag{B48}$$

Clearly Δ_4' and Δ_4''' are limiting cases of Δ_4''.

At the other end of the spectrum we might assert diffusion is altogether negligible, H_{15}. In this case the flat profile at $t=0$ becomes a paraboloid whose tip advances with twice the mean speed. For this

$$\Delta_5 \quad \bar{c} = \begin{cases} 1 - x/2Ut, & x < 2Ut, \\ 0, & x > 2Ut. \end{cases} \tag{B49}$$

The correlation of results from a model.

We have already mentioned that an "exact apparent" diffusion coefficient can be fitted to the results of the calculation with Δ_2 so that in Δ_4 it reproduces the same mean concentration. Unless some serendipitous constancy emerges this exercise serves only to compare the models for the particular problem. Another method is to look for a correlation between the results of one model and those of another without attempting to give significance to the constants. This is again limited in usefulness to a particular problem and is an empirical exercise with interpolatory value but no extrapolatory or explanatory power. For example Gill et al. [74] found that a good approximation to the solution of Δ_2 was afforded by the formula

$$\bar{c} = \frac{1}{2} \operatorname{erfc} \frac{x - Ut}{2\sqrt{Dt\,(1+P^2 Q)}} + \frac{1}{2} \exp \frac{xU}{D(1+P^2 Q)} \operatorname{erfc} \frac{x + Ut}{2\sqrt{Dt\,(1+P^2 Q)}} \tag{B50}$$

where, as before, $P = aU/D$ and

$$\Delta_6: \quad Q = 0.028(Dt/a^2)^{0.55}, \quad Dt/a^2 < 0.6 \tag{B51}$$

148

It is worth noting that Δ_4'' corresponds to Δ_5 but with $Q = 1/48$, while Δ_4''' would correspond to $Q = 0$.

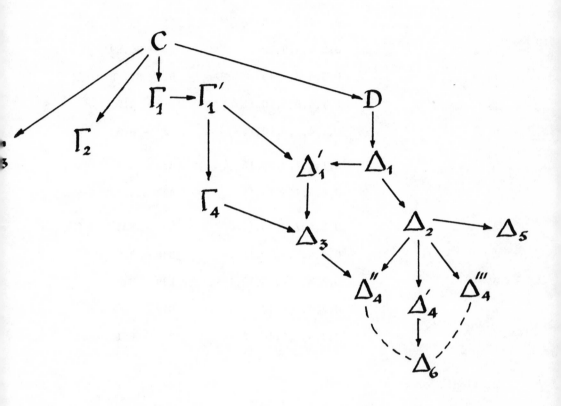

Summary.

Model	Hypotheses	Equations
Γ_1	0,1,4,8,9,10	B1 - B6
Γ_1'	0,1,4,8,9,10	B1 - B8
Γ_2	0,2,4,8,9,10	B1 - B5,B9
Γ_3	0,3,4,8,9,10	B1 - B5, B10,B11
Γ_4	0,1,4,8,9,10,11	B12 - B19
Δ_1	0,1,4,7,9,10,14	B22 - B25
Δ_1'	0,1,4,7,9,10,14	B23 - B27
Δ_2	0,2,4,7,9,10,14	B22,B23,B28 - B30
Δ_2'	0,2,4,7,9,10,14	B31 - B35
Δ_3	0,2,4,7,9,10,11,14	B36 - B38
Δ_4	0,2,4,7,9,10,12,14	B39 - B43
Δ_5	0,2,4,7,9,10,14,15	B49
Δ_6	0,2,4,7,9,10,12,14	B50 - B51

Notation.

A_i	area of ith phase $\pi(r_i^2 - r_{i-1}^2)$
a	radius of tube, common value of r_1 and r_2
c	concentration of solute in tube (system D)
c_i	concentration of solute in i^{th} phase
c_{if}	feed concentration
$c_i^{(p)}, c^{(p)}$	p^{th} moment of concentration
$\bar{c}$	average concentration
$<c>$	cup mixing mean concentration
D_e	equivalent longitudinal dispersion coefficient

D_i	diffusion coefficient in i^{th} phase
k	rate constant for partition
L	length of tube
$m^{(p)}$	p^{th} moment of mean concentration
P	Peclet number aU/D
Q	defined in eqn. (B51)
r	radial distance ($r_0 = 0$; r_1 interface of phases; r_2, outer radius of phase 2)
t	time
U_i	mean velocity of phase i
V	velocity of moving origin
x	length coordinate
y	x-Vt
α	partition coefficient
β_i	$A_i/(A_1 + A_2)$
κ	constants in eqn. (B21)
ξ	$xD^{1/2}/2Ut^{1/2}$
ρ	r/a
τ	Dt/a^2
ϕ_i	velocity distribution
χ_i	initial distribution
$\chi_i^{(p)}$	p^{th} moment of initial distribution

Appendix C The stirred tank reactor

A stirred tank reactor consists of a cylindrical vessel of volume V with incoming and outgoing pipes. The incoming pipes bring reactants A_1, A_2...A_r, at volume flow rates q_1, q_2,...q_r, and the outgoing pipe takes of the mixture of products A_{r+1},...A_s and the remnants of the reactants, at a flow rate of $q = q_1 + q_2 + ... + q_r$. Thus the volume V remains constant. The reaction can be written as $\Sigma \alpha_j A_j = 0$, where α_{r+1},...α_s are

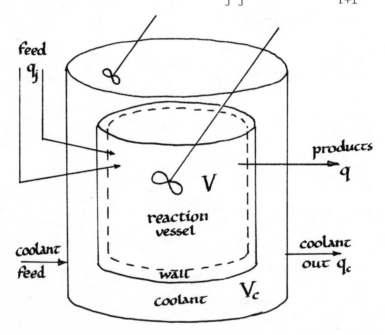

positive. This cylinder is immersed in another cylinder of annular volume V_c, also perfectly stirred, which is fed cooling water of temperature T_{cf} at a flow rate of q_c. Other details will be mentioned as we proceed. It should be mentioned that this description has already been deliberately simplified since the geometry of a real jacketed reactor would undoubtedly be more complicated than that of simple cylinders. However, I have no desire to pile Pelion on Ossa.

Hypotheses.

Let us lump together the applicability of all physical laws, such as the conservation of matter and energy or Fourier's law of heat conduction and call this the underlying general hypothesis, H_o. The following hypotheses can be extracted from the description or be excogitated as relevant to the setting:

H_1: the mixing is perfect so that the concentrations c_j, the reaction temperature T and T_c, the temperature of the coolant jacket, are all independent of position, though they may be functions of time. The volumes V and V_c are constant, as also are the flow rates q_j and the feed temperatures T_{jf}. The work done by the stirrers may be ignored.

H_2: the reaction rate is a function $r(c_1, \ldots c_s, T)$ such that the rate of change in the number of moles of A_j by reaction alone is $\alpha_j r$ per unit volume.

H_3: the heat transfer to the inner and outer sides of the wall where the surface temperatures will be denoted by T_i and T_o respectively can be described by transfer coefficients h_i and h_o such that the heat transferred per unit area is $h_i(T-T_i)$ and $h_o(T_o-T_c)$ respectively.

H_4: the heat capacity of the reaction mixture does not change significantly.

H_5: the system is in steady state.

H_6: the curvature of the wall is negligible and the sharp corners can be ignored.

153

H_7:   the conductivity of the wall is extremely high.

H_8:   the heat capacity of the wall is negligible.

H_9:   the response of the cooling jacket is virtually instantaneous.

H_{10}: the reaction is the first-order and irreversible with respect to

the key species.

Derivation of the most general model Σ_1.
_____

Using the principle of the conservation of matter from the underlying

hypothesis H_o, we have the following balance for each species:

$$
\begin{bmatrix} \text{rate of change of} \\ \text{number of moles of} \\ A_j \text{ in reactor} \end{bmatrix}
=
\begin{bmatrix} \text{rate of} \\ \text{feed of} \\ A_j \end{bmatrix}
-
\begin{bmatrix} \text{rate of} \\ \text{withdrawal} \\ \text{of } A_j \end{bmatrix}
+
\begin{bmatrix} \text{rate of formation} \\ \text{of } A_j \text{ by means of} \\ \text{the reaction} \end{bmatrix}
$$

If c_{jf} is the concentration (moles/volume) of A_j in its feed stream,

this translates immediately into the ordinary differential equation

$$V \frac{dc_j}{dt} = q_j c_{jf} - q c_j + \alpha_j V r(c_1,\ldots c_s,T). \tag{C1}$$

In obtaining this equation we have invoked H_1 and H_2.

If $h_j(c_1,\ldots c_s,T)$ is the enthalpy per mole of A_j, and the work done by

the stirrer is ignored, then conservation of energy implies

$$V \frac{d}{dt} \Sigma c_j h_j = \Sigma q_j c_f h_{jf} - q \Sigma c_j h_j - A_i h_i (T - \bar{T}_i)$$

The symbol h_{jf} denotes the specific enthalpy of A_j evaluated for its

feed conditions. In the last term A_i is the total internal wall area and

since the heat transfer coefficient h_i is independent of position we need

only average the inner surface temperature of the wall. Thus H_1 and H_3

are used here. We now simplify this equation by subtracting from it the sum

over j of equations (C1) each multiplied by h_j. Thus

$$V\Sigma_j c_j \frac{dh_j}{dt} = q_j c_{jf}(h_{jf}-h_j) - (\Sigma\alpha_j h_j)V_r - A_i h_i(T-\bar{T}_i)$$

Next we observe that $\alpha_j h_j = \Delta H$ is the heat of reaction and that since h_j is an intensive thermodynamic variable (i.e. $\Sigma c_j \frac{\partial h_j}{\partial c_k} = \Sigma c_j \frac{\partial h_k}{\partial c_j} = 0$)

$$\Sigma c_j \frac{dh_j}{dt} = \Sigma c_j c_{pj} \frac{dT}{dt} + \Sigma\Sigma c_j \frac{\partial h_j}{\partial c_k} \frac{dc_k}{dt} = C_p \frac{dT}{dt}$$

where c_{pj} is the heat capacity of A per mole and C_p is the heat capacity of the mixture per unit volume. We now invoke H_4 and write $q_j c_{jf}(h_{jf}-h_j) = qC_p(T_f-T)$ to give

$$VC_p \frac{dT}{dt} = qC_p(T_f-F) + (-\Delta H)Vr(c_1,\ldots T)-A_i h_i(T-\bar{T}_i). \qquad (C2)$$

This form allows us to check the common sense of the equation for we can write it as:

$$\begin{bmatrix}\text{rate of}\\ \text{change}\\ \text{of heat}\\ \text{content}\end{bmatrix} = \begin{bmatrix}\text{heat}\\ \text{brought}\\ \text{in with}\\ \text{feed}\end{bmatrix} - \begin{bmatrix}\text{heat}\\ \text{taken out}\\ \text{with}\\ \text{products}\end{bmatrix} + \begin{bmatrix}\text{heat}\\ \text{generated}\\ \text{by}\\ \text{reaction}\end{bmatrix} - \begin{bmatrix}\text{heat}\\ \text{removed by}\\ \text{cooling}\\ \text{wall}\end{bmatrix}$$

The wall has been simplified to be a finite cylinder of internal area, A_i. If we denote the region it occupies by D and its inner and outer surfaces by ∂D_i and ∂D_o, we apply conservation principles and Fourier's law of heat conduction to obtain for the wall temperature T_w,

$$\rho_w c_{pw} \frac{\partial T_w}{\partial t} = k_w \nabla^2 T_w \quad \text{in } D, \qquad (C3)$$

where ρ_w, c_{pw} and k_w are the density, specific heat and conductivity of the wall respectively. To obtain boundary conditions we have to call on H_3,

$$k_w \frac{\partial T_w}{\partial n} = h_i(T-T_i) \quad \text{on } \partial D_i, \qquad (C4)$$

155

$$k_w \frac{\partial T_w}{\partial n} = h_o(T_c - T_o) \quad \text{on } \partial D_o, \tag{C5}$$

where $\partial/\partial n$ is the normal derivative on the surface directed outward from D.

Before seeing how these equations simplify let us write the heat balance equation for the coolant. This is

$$V_c C_{pc} \frac{dT_c}{dt} = q_c C_{pc}(T_{cf} - T_c) + A_o h_o(\bar{T}_o - T_c) \tag{C6}$$

where C_{pc} is the heat capacity of the coolant per unit volume, A_o is the area of ∂D_o and $\bar{T}_o$ the average outer temperature. H_1 and H_3 have been involved in deriving this equation as well as the underlying H_o.

Equations (C1-6), together with suitable initial conditions, give $s+2$ ordinary and one partial differential equation with its boundary conditions and constitute Σ_1, the most detailed model we shall consider. In obtaining it the hypothesis used have been H_o, H_1, H_2, H_3 and H_4.

Derivation of the steady state models Σ_2 and Σ_3.

Now let us invoke H_5 and assume that the system is at steady state. To do this is to set all time derivatives equal to zero and it leads to a partial differential equation, Laplace's, for T_w connected through its boundary conditions to a set of algebraic equations. (The term algebraic equation is applied to any equation that is not a differential equation even though transcendental functions may appear in it.) Let this model be Σ_2.

However when we recognize that T_w is a potential function we can use Green's theorem to give

$$0 = \iiint_D k_w \nabla^2 T_w dV = \iint_{\partial D_i + \partial D_o} k_w(\partial T_w/\partial n)dS$$

$$= A_i h_i(T - \bar{T}_i) - A_o h_o(\bar{T}_o - T_c) \tag{C7}$$

Combining this with (C6) we have three expressions for the rate of removal of heat

$$Q_c = q_c C_{pc}(T_c - T_{cf}) = A_o h_o(\bar{T}_o - T_c) = A_i h_i(T - \bar{T}_i).$$

This gives

$$T - T_{cf} = Q_c \left\{ \frac{1}{q_c C_{pc}} + \frac{1}{A_o h_o} + \frac{1}{A_i h_i} + \frac{\bar{T}_i - \bar{T}_o}{Q_c} \right\}$$

but the last term is a messy one and has to be evaluated from the full solution of the potential equation. If however we invoke H_6 and take d_w to be the thickness of the wall then we have the local flux of heat per unit area equal to $k_w(T_i - T_o)/d_w$. Also ignoring curvature makes $A_o = A_i = A$ so that $Q_c = k_w A(T_i - T_o)/d_w$. Then

$$Q_c = \hat{h}A(T - T_{cf}) \quad \text{where} \quad \frac{1}{\hat{h}} = \frac{A}{q_c C_{pc}} + \frac{1}{h_o} + \frac{1}{h_i} + \frac{d_w}{k_w} \qquad (C8)$$

Thus the model simplifies to Σ_3 consisting of the equations

$$q_j c_{jf} - q c_j + \alpha_j Vr(c_1, \ldots c_s, T) = 0, \quad j = 1, \ldots s \qquad (C9)$$

$$q C_p(T_f - T) + (-\Delta H)Vr(C_1, \ldots c_s, T) - \hat{h}A(T - T_{cf}) = 0 \qquad (C10)$$

Notice that H_8 has no relevance at all to this model and that were we to invoke H_7 it would not change the model but only modify the value of $\hat{h}$.

Simplified transient models Σ_4 and Σ_5.

Let us return now to the transient model, dropping H_5, and see the effect of H_7 and H_8. A little caution is needed here or terms can get lost. First suppose that $k_w \to \infty$ (H_7), we cannot conclude that T_w is constant in time by observing that this limit, like the steady state hypothesis, leads to Laplace's equation for T_w. Rather the form of the temperature profile through the wall is invariant in time, for it takes up a temperature profile

157

between T_i and T_o with virtually no delay. Thus, for very large k_w, the Laplacian of the temperature in the wall becomes very small and the product $k_w \nabla^2 T_w$ is finite.

Let T_w denote the mean temperature of the wall, which is uniform with respect to position around the reactor on account of $k_w \rightarrow \infty$ and the fact that the wall is exposed to uniform temperatures on both sides. Then integrating (C3) throughout D, using Green's theorem and the boundary conditions (C4) and (C5), gives

$$V_w \rho_w C_{pw} \frac{dT_w}{dt} = A_i h_i T + A_o h_o T_c - (A_i h_i + A_o h_o) T_w \tag{C11}$$

Thus we have a model consisting of (S+3) ordinary differential equations (C1), (C2), (C6) and (C11), where in (C2) and (C6), T_i and T_o have both been made equal to T_w. Let this be the model Σ_4; it invokes H_o-H_4 and H_7. Note that we did not need H_8.

Suppose now that we assert H_8, but drop H_7. Again T_w is governed by Laplace's equation and we can arrive at (C7). However we are still left with Laplace's equation unless we invoke H_6 and note that

$$Q_c = A h_o (\bar{T}_o - T_c) = A k_w (\bar{T}_i - \bar{T}_o)/d_w = A h_i (T - \bar{T}_i) = A h^*(T - T_c) \tag{C12}$$

where

$$\frac{1}{h^*} = \frac{1}{h_o} + \frac{d_w}{k_w} + \frac{1}{h_i}$$

We then have a model Σ_5, consisting of (S+2) ordinary equations; the S equation (C1) and

$$VC_p \frac{dT}{dt} = qC_p(T_f - T) + (-\Delta H)Vr(c_1, \ldots T) - A h^*(T - T_c) \tag{C13}$$

$$V_c C_{pc} \frac{dT_c}{dt} = q_c C_{pc}(T_{cf} - T_c) + A h^*(T - T_c) \tag{C14}$$

158

Finally two more hypotheses are introduced to give a specially important case. The first, H_9, is of the nature of the limiting hypotheses we have been making. It asserts that $V_c/q_c \ll V/q$ so that, by comparison with the other time derivatives, dT_c/dt has a small multiplier. If we go to the limit and say that the response of the cooling jacket is virtually instantaneous, we wipe out the derivative in eqn. (C6). Thus we are essentially back in the steady state and can write $Q_c = hA(T-T_{cf})$ as in eqn. (C8).

The second hypothesis, H_{10}, asserts that the reaction is irreversible and first order in the concentration of one species. If this is A_1 we can write $\alpha_1 = -1$ and $r = kc_1$, where $k = k(T)$ is a function only of T. For simplicity we can then drop the suffix on c and eqns. (C1) and (C2) become

$$V \frac{dc}{dt} = q(c_f - c) - Vk(T)c \tag{C15}$$

$$VC_p \frac{dT}{dt} = qC_p(T_f - T) + (-\Delta H)Vk(T)c - hA(T-T_{cf}) \tag{C16}$$

These equations give a pair of equations for a pair of unknowns $c(t)$, $T(t)$.

The dimensionless equations.

Up to this point everything has been very dimensional and it is not clear what we have mean by large and small values. There are various characteristic lengths, times etc. in the problem and we want to pick the most judicious set. In particular constants to which we are going to give some limiting value should not be used to render others dimensionless, nor should those whose variation we are going to study.

Let V/q be the characteristic time and $\tau = qt/V$;

$c*$, be a characteristic concentration, say $\Sigma q_j c_{jf}/q$ and $u_j = c_j/c*$;

$T*$, be a characteristic temperature, say $(-\Delta H)c*/C_p$, and $v = T/T*$, $w = T_w/T*$, $\Theta = T_c/T*$, $v_f = T_f/T*$, $\Theta_f = T_{cf}/T*$, $w_i = T_i/T*$, $w_o = T_o/T*$;

d_w, be a characteristic length such as a mean wall thickness with which the independent variables in the Laplacian are to be made dimensionless.

The other parameters will emerge with the equations. If (C1) is divided by $qc*$ it becomes

$$\frac{du_j}{d\tau} = \gamma_j - u_j + \alpha_j R(u_1, \ldots, v) \tag{C18}$$

where $\gamma_j = q_j c_{jf}/\Sigma q_j c_{jf}$, i.e. $\Sigma \gamma_j = 1$, is the jth fraction of feed and $R = Vr/qc*$. Similarly let (C2) be divided by $qC_p T*$ to give

$$\frac{dv}{d\tau} = v_f - v + R(u_1, \ldots v) - \beta_i(v - v_i) \tag{C19}$$

where $\beta_i = A_i h_i/qC_p$.

We will use the same symbol as before for the Laplacian with respect to the dimensionless variables, so that (C3) can be divided by $\rho_w c_{pw} q/VT*$ to give

$$\frac{\partial w}{\partial \tau} = \lambda \nabla^2 w \tag{C20}$$

where $\lambda = k_w V/q\rho_w c_{pw} d_w^2$. If $\partial/\partial\nu$ denotes the normal derivative in the dimensionless variables

$$\lambda \frac{\partial w}{\partial \nu} = \frac{\beta_i}{\delta_w}(v - w_i) \text{ on } \partial D_i \tag{C21}$$

where $\delta_w = A_i d_w \rho_w c_{pw}/VC_p$ is the ratio of the heat capacity of the wall to that of the contents. Similarly

160

$$\lambda \frac{\partial w}{\partial \nu} = \frac{\beta_o}{\delta'_w} (\Theta - w_o) \text{ on } \partial D_o \tag{C22}$$

where $\beta_o = A_o h_o / qC_p$ and $\delta'_w = (A_o/A_i)\delta_w$. Finally (C5) becomes

$$\delta_c \frac{d\Theta}{d\tau} = \chi(\Theta_f - \Theta) + \beta_o(w_o - \Theta) \tag{C23}$$

where $\delta_c = V_c C_{pc}/VC_p$ and $\chi = q_c C_{pc}/qC_p$.

In the later models we have $\hat{h}$ and $h*$ and we make them dimensionless with qC_p to give

$$\hat{\beta} = A\hat{h}/qC_p = \left\{ \frac{1}{\chi} + \frac{1}{\beta_o} + \frac{1}{\beta_i} + \frac{1}{\lambda\delta_w} \right\}^{-1} \tag{C24}$$

$$\beta* = Ah*/qC_p = \left\{ \frac{1}{\beta_o} + \frac{1}{\lambda\delta_w} + \frac{1}{\beta_i} \right\}^{-1} \tag{C25}$$

Thus we have for Σ_3 the (S+1) non-differential equations

$$\gamma_j - u_j + \alpha_j R(u_1, \ldots v) = 0, \tag{C26}$$

$$v_f - v + R(u_1, \ldots v) - \hat{\beta}(v - \Theta_f) = 0 \tag{C27}$$

The model Σ_4 consists in eqns. (C18), (C19) and (C23) with the dimensionless form of (C11), namely

$$\delta_w \frac{dw}{d\tau} = \beta_i v + \beta_o \Theta - (\beta_o + \beta_i)w. \tag{C28}$$

Finally, the model Σ_5 in eqn. (C15) and the two equations

$$\frac{dv}{d\tau} = v_f - v + R(u_1, \ldots v) - \beta*(v - \Theta) \tag{C29}$$

$$\delta_c \frac{d\Theta}{d\tau} = \chi(\Theta_f - \Theta) + \beta*(v - \Theta) \tag{C30}$$

The initial conditions, as needed, are

$$u_j = u_{jo}, v = v_o, w = w_o(\underset{\sim}{\xi}), \Theta = \Theta_o, \tau = 0, \underset{\sim}{\xi} = \underset{\sim}{x}/d_w \tag{C31}$$

The ways in which Σ_6 can be non-dimensionalized are discussed extensively in Sec. 4.2.

161

Summary of parameters.

Reaction: α_j stoichiometric coefficients

        —   parameters of the rate law e.g. E/RT_f

Feed: γ_j fraction of A_j in feed

    v_f feed temperature

    Θ_f coolant feed temperature

Capacities: δ_c heat capacity ratio of coolant to reactants

    δ_w ratio of heat capacity of wall to reactants

    χ ratio of heat carrying capacities of coolants

    to reactants

Transfer: β_o,β_i dimensionless heat transfer coefficients

    $\hat{\beta},\beta*$ composite heat transfer-coefficients

    λ dimensionless wall conductivity

Summary of models.

Model Σ	Hypotheses H	Equations C	Dimensionless equations	Remarks
1	0,1,2,3,4	1,2,3,4,5,6	18,19,20,21,22,23	
2	0,1,2,3,4,5	1,2,3,4,5,6	18,19,20,21,22,23	Set $\partial/\partial t$ or $\partial/\partial\tau=0$
3	0,1,2,3,4,5,6	9,10	26,27	
4	0,1,2,3,4,7	1,2,6,11	18,19,23,28	
5	0,1,2,3,4,6,8	1,13,14	18,29,30	
6	0,1,2,3,4,6,8,9,10	15,16	See Sec. 4.2	

Notation for Appendix C: the system S and its models Σ.

A_j chemical species, $j = 1, S$; $j = 1...r$ for reactants, $r+1,...S$

    products

A_i, A_o inner and outer areas of reactor wall

C_p heat capacity per unit volume of reaction mixture

162

C_{pc} heat capacity per unit volume of coolant

c_j concentration of A_j

c_{jo} initial concentration of A_j

c_{jf} feed concentration of A_j

c_{pw} specific heat of wall

c^* reference concentration

d_w thickness of reactor wall

h_j enthalpy per mole of A_j

h_{jf} enthalpy per mole of A_j under feed conditions

h_i, h_o heat transfer coefficient at inner and outer wall surfaces

$\hat{h}, h^*$ composite heat transfer coefficients

k_w thermal conductivity of wall

n outward normal to wall in $\partial/\partial n$

Q_c total rate of heat removal

q flow rate of reacting mixture

q_c coolant rate flow

q_j feed rate of A_j

R dimensionless reaction rate Vr/qc^*

r reaction rate per unit volume

S number of reacting species

T temperature

$T_c, T_{cf}, T_f, T_w(x)$ temperature of coolant, coolant feed, reactor feed and wall resp.

$T_i, T_o, \bar{T}_i, \bar{T}_o$ inner and outer wall temperatures and their averages

T_o', T_{wo}, T_o initial reactor, wall and coolant temperatures

T^* reference temperature

t time

V	volume of reactor
V_c, V_w	volume of coolant, wall
v	dimensionless temperature T/T^*
v_f	T_f/R^*
$w(\xi)$	T_w/T^*
w_i/w_o	$T_i/T^*, T_o/T^*$
$\underset{\sim}{x}$	coordinates within the wall
α_j	stoichiometric coefficients
$\hat{\beta}_i, \beta_o$	$h_i A_i/qC_p, h_o A_o/qC_p$
β, β^*	dimensionless composite heat transfer coefficients; (C24), (C25)
γ_j	dimensionless feed rate of A_j
$\delta_c, \delta_w, \delta_w'$	$V_c C_{pc}/VC_p, A_i d_w \rho_w c_{pw}/VC, A_o d_w \rho_w c_{pw}/VC_p$
ΔH	heat of reaction
Θ	dimensionless coolant temperature, T_c/T^*
Θ_o, Θ_f	$T_{co}/T^*, T_{cf}/T^*$
λ	$k_w V/q_w c_{pw} d_w^2$
ν	dimensionless normal in $\partial/\partial\nu$
$\underset{\sim}{\xi}$	x/d_w
ρ_w	density of wall
τ	qt/V
χ	$q_c C_{pc}/qC_p$

164

References

1   P. Achinstein   Models, analogies and theories. Phil. of
                                     Sci. 31 (1964) 328.

2   P. Achinstein   Theoretical models. Brit. Jnl. for the
                                     Phil. of Sci. 16 (1965) 102.

3   R. Ackermann   Confirmatory models of theories. Brit.
                                     Jnl. for the Phil. of Sci. 16 (1966) 312.

4   W. P. Alston   Philosophy of language. (Englewood
                                     Cliffs: Prentice Hall, 1964).

5   N. R. Amundson and R. Aris   Some remarks on longitudinal diffusion
                                     or mixing in fixed beds. A.I.Ch.E. J. 3
                                     (1957) 280.

6   N. R. Amundson and S-L Liu   Stability of adiabatic packed bed
                                     reactors. A simplified treatment.
                                     I.E.C. Fundamentals 1 (1962) 200.

7   L. Apostel   Towards the formal study of models in
                                     the non-formal sciences. In [68] p. 1.

8   M. A. Arbib   Theories of abstract automata.
                                     (Englewood Cliffs: Prentice Hall, 1968).

9   M. A. Arbib and E. G. Manes   The categorical imperative: arrows,
                                     structures and functors. (New York:
                                     Academic Press, 1975).

10 R. Aris   On the dispersion of a solute in a
                                     fluid flowing through a tube. Proc.
                                     Roy. Soc. A235 (1956) 67.

11 R. Aris   On the dispersion of a solute by
                                     diffusion, convection and exchange
                                     between phases. Proc. Roy. Soc. A252
                                     (1959) 538.

12 R. Aris   Some problems in the analysis of tran-
                                     sient behavior and stability of chemical
                                     reactors. Adv. in Chem. 109 (1972) 578.

13 R. Aris   On the ostensible steady state of a
                                     dynamical system. Rend. Lincei. Sr.
                                     VIII 57 (1974) 1.

14 R. Aris   The mathematical theory of diffusion
                                     and reaction in permeable catalysts.
                                     (Oxford: Clarendon Press, 1975) 2 vols.

15   R. Aris   How to get the most out of an equation
                                       without really trying. Chem. Eng. Educ.
                                       10 (1976) 114.

16   R. Aris and A. E. Humphrey   The dynamics of a chemostat in which
                                       two organisms compete for a common
                                       substrate. Biotech. and Bioeng. 19
                                       (1977) 1375.

17   R. Aris and D. L. Schruben   Transients in distributed chemical
                                       reactors. Part I. A simplified model.
                                       Chem. Eng. J. 2 (1972) 179.

18   R. Aris and S. Viswanathan   An analysis of the countercurrent movin
                                       bed reactor. SIAM/AMS Proceedings 8
                                       (1974) 99.

19   A. M. Arthurs   Complimentary variational principles.
                                       (Oxford: Clarendon Press, 1970).

20   P. Auger   Models in science. Diogenes 52 (1965)
                                       1.

21   J. E. Bailey   Lumping analysis of reactions in con-
                                       tinuous mixtures. Chem. Eng. J. 3
                                       (1972) 52.

22   M. S. Bartlett   Introduction to stochastic processes.
                                       (Cambridge: Cambridge Univ. Press, 1966

23   J. W. L. Beament (Ed.)   Models and analogues in biology (Sympos
                                       of the Society for Experimental Biology
                                       No. 14). (Cambridge: Cambridge Univ.
                                       Press, 1960).

24   R. Bellman, K. L. Cooke and J. A. Lockett, Algorithms, graphs and
                                       computers. (New York: Academic Press,
                                       1970).

25   R. Bellman and M. Giertz   On the analytic formalism of the theor
                                       of fuzzy sets. Information Sciences 5
                                       (1973) 149.

26   R. Bellman and L. A. Zadeh   Decision-making in a fuzzy environment
                                       Management Sciences 17 (1970) 8141.

27   C. Berge   The theory of graphs and its applica-
                                       tions. (London: Methuen, 1964).

28   C. Berge and A. Ghouila-Houri   Programming, games and transportation
                                       networks. (London: Methuen, 1965).

29   J. M. Beshers   Models and theory construction. In
                                       M. L. Barron (ed.)(1966) 590.

30   K. B. Bischoff   An extension of the general criterion for the importance of pore diffusion. Chem. Eng. Sci. 22 (1967) 525.

31   J. T. Bonner   Analogies in biology. Synthese 15 (1963) 275.

32   K. C. Bowen   Mathematical battles. Bull. I.M.A. 9 (1973) 310.

33   R. B. Braithwaite   Scientific explanation. A study of the function of theory, probability and law in science. (Cambridge: Cambridge Univ. Press, 1953).

34   R. B. Braithwaite   Models in the empirical sciences. In Nagel, E. et al. (ed.) (1962) 224.

35   M. Brodbeck   Models, meanings and theories. In [77] (1959) 373.

36   M. Brodbeck   Models, meaning and theories. In Symposium on Sociological Theory (ed. L. Gross)(New York: Harper and Row, 1959). (Also in Readings in the philosophy of the social sciences (ed. M. Brodbeck)(New York: Macmillan, 1968).

37   M. Bunge   Models in theoretical science. Akten des XIV Int. Kong. für Philosophie (Herber, Wien)(1968) 208.

38   R. R. Bush and F. Mosteller   A comparison of eight models. In R. R. Bush and W. K. Estes (eds.) Studies in mathematical learning theory. (Stanford: Stanford Univ. Press, 1961). Also in [111].

39   K. V. Bury   Statistical models in applied science. (New York: John Wiley, 1975).

40   H. Byerly   Model structures and model objects. Brit. Jnl. for the Phil. of Sci. 20 (1969) 135.

41   N. R. Campbell   The foundations of science; the philosophy of theory and experiment. (New York: Dover, 1957).

42   P. C. Chatwin   The approach to normality of the concentration distribution of a solute in a solvent flowing along a straight pipe. J. Fluid Mech. 43 (1970) 321.

167

43 R. F. Churchhouse   Discoveries in number theory aided by computers. Bull. I.M.A. 9 (1973) 15.

44 W. A. Coppel   Stability and asymptotic behavior of differential equations. (Boston: Heath, 1965).

45 C. A. Coulson   Mathematics and the real world. Bull. I.M.A. 9 (1973) 2.

46 C. A. Coulson   The rôle of mathematics in chemistry. Bull. I.M.A. 9 (1973) 206.

47 C. A. Coulson   Mathematical models. Bull. I.M.A. 10 (1974) 340.

48 D. R. Cox and H. O. Miller The theory of stochastic processes. (New York: John Wiley, 1965).

49 J. Crank   Diffusion mathematics in medicine and biology. Bull. I.M.A. 12 (1976) 106.

50 J. Crank and R. D. Prahle Melting ice by the isotherm migration method. Bull. I.M.A. 9 (1973) 12.

51 I. Dambska   Modèle et objet de la connaissance. Revue Intnl. de Phil 87 (1969) 34.

52 G. B. Dantzig   Linear programming and extensions. (Princeton: Princeton Univ. Press, 1963)

53 H. A. Deans and L. Lapidus A computational model for predicting and correlating the behavior of fixed bed reactors. A.I.Ch.E. J. 6 (1960) 656.

54 D. J. De Solla Price   Automata and the origins of mechanism and mechanistic philosophy. Tech. and Culture 5 (1964) 5.

55 K. W. Deutsch   Mechanism, organism and society: some models in natural and social science. Phil. of Sci. 18 (1951) 230.

56 P. Duhem   The aim and structure of physical theory (Trs. P. P. Wiener). (1st Edn. 1906). (Princeton: Princeton Univ. Press, 1954)

57 A. W. F. Edwards   Models in genetics. In [23] p. 6.

58 P. Edwards (ed.)   Encyclopedia of Philosophy. (London, 1967). (8 vols.)

59   G. Eigenberger   On the dynamic behavior of the catalytic fixed-bed reactor in the region of multiple steady states--I. The influence of heat conduction in two phase models. Chem. Eng. Sci. 27 (1972) 1909.

60   G. Eigenberger   On the dynamic behavior of the catalytic fixed-bed reactor in the region of multiple steady states--II. The influence of the boundary conditions in the catalyst phase. Chem. Eng. Sci. 27 (1972) 1917.

61   G. Eigenberger   Kinetic instabilities in catalytic reactions--a modelling approach. Proc. 4th Int. Symp. on Chem. Reaction Eng. Heidelberg, April, 1976. Dechema Frankfurt.

62   S. Eilenberg   Automata, languages and machines. (New York: Academic Press, 1974).

63   G. L. Farre   Remarks on Swanson's theory of models. Brit. Jnl. for the Phil. of Sci. 18 (1967) 140.

64   W. Feller   Introduction to probability theory and its applications (New York: John Wiley, 1968) 2 vols.

65   P. C. Fife   Pattern formation in reacting and diffusing systems. J. Chem. Phys. 64 (1976) 554.

66   B. A. Finlayson   The method of weighted residuals and variational principles. (New York: Academic Press, 1972).

67   L. Ford and D. Fulkerson   Flows in networks. (Princeton: Princeton Univ. Press, 1962).

68   H. Freudenthal   The concept and the role of the model in mathematics and social sciences. (Dordrecht: Reidel Pub. Co., 1961).

69   A. Friedman   Differential games. (New York: Wiley-Interscience, 1971).

70   H. L. Frisch   Time lag in transport theory. J. Chem. Phys. 36 (1962) 510.

71   H. L. Frisch   The time lag in diffusion. J. Phys. Chem. 61 (1957) 93.

72   G. R. Gavalas   Nonlinear differential equations of
                                 chemically reacting systems.
                                 (Heidelberg: Springer Verlag, 1968).

73   G. Gavalas and R. Aris   On the theory of reactions in continuous
                                 mixtures. Phil. Trans. Roy. Soc. A260
                                 (1966) 351.

74   W. N. Gill, V. Ananthakrishnan and H. J. Barduhn, Laminar dispersion
                                 in capillaries. A.I.Ch.E. J. 11 (1965)
                                 1063.

75   W. N. Gill and R. J. Nunge   Mechanisms affecting dispersion and
                                 miscible displacement. Ind. Eng. Chem.
                                 61 (Pt. 9)(1969) 33.

76   H. J. Groenewold   The model in physics. In [68] p. 68.

77   L. Gross (ed.)   Symposium on sociological theory.
                                 (Evanston: Northwestern Univ. Press,
                                 1959).

78   M. Gross   Mathematical models in linguistics.
                                 (Englewood Cliffs: Prentice-Hall, 1972).

79   G. G. Hall   Modelling--a philosophy for applied
                                 mathematicians. Bull. I.M.A. 8 (1972)
                                 226.

80   J. M. Hammersley   Maxims for manipulators. Bull. I.M.A.
                                 9 (1973) 276; 10 (1973) 368.

81   J. M. Hammersley   How is research done? Bull I.M.A. 9
                                 (1973) 214.

82   J. M. Hammersley   Poking about for vital juices of
                                 mathematical research. Bull. I.M.A.
                                 10 (1974) 235.

83   F. Harary   'Cosi fan Tutte'--a structural study.
                                 Psych. Reports 13 (1963) 466.

84   F. Harary   Graph theory. (Reading: Addison-
                                 Wesley, 1969).

85   F. Harary, F. R. Norman and D. Cartwright, Structural models: an
                                 introduction to the theory of directed
                                 graphs. (New York: John Wiley, 1965).

86   R. Harré   An introduction to the logic of the
                                 sciences. (London, 1960).

87   R. Harré   The principles of scientific thinking.
                                    (London, 1970).

88   N. Hawkes (ed.)   International seminar on trends in
                                    mathematical modelling. (Lec. Notes
                                    in Econ. and Math. Systems 80).
                                    (Heidelberg: Springer Verlag, 1973).

89   M. B. Hesse   Operational definition and analogy in
                                    physical theories. Brit. Jnl. for the
                                    Phil. of Sci. 2 (1952) 281.

90   M. B. Hesse   Models in physics. Brit. Jnl. for the
                                    Phil. of Sci. 4 (1954) 198.

91   M. B. Hesse   Science and the human imagination:
                                    aspects of the history and logic of
                                    physical science. (London, 1954).

92   M. B. Hesse   Models and analogies in science.
                                    (London: Sheed and Ward, 1963).

93   M. B. Hesse   Models and analogy in science. In
                                    P. Edwards. (ed.)(1967) Vol. V.

94   M. Hesse   The structure of scientific inference.
                                    (Berkeley: Univ. of Cal. Press, 1974).

95   V. Hlaváĉek, M. Marek and M. Kubíĉek, Analysis of nonstationary heat
                                    and mass transfer in a porous catalyst
                                    particle. J. Cat. 15 (1969) 17, 31.

96   F. R. Hodson, D. G. Kendall and P. Tautu, Mathematics in the
                                    archaeological and historical sciences.
                                    (Edinburgh: Univ. Press, 1971).

97   E. H. Hutten   The role of models in physics. Brit.
                                    Jnl. for the Phil. of Sci. 4 (1954) 284.

98   E. H. Hutten   The ideas of physics. (London, 1967).

99   R. Issacs   Differential games. (New York: John
                                    Wiley, 1965).

100 K. Itô   On stochastic differential equations.
                                    Mem. Amer. Math. Soc. No. 4 (1951).

101 R. E. Kalman   On the mathematics of model building.
                                    In E. R. Caianello (ed.) Neural Networks.
                                    (Heidelberg: Springer Verlag, 1968).

102 A. Kaplan   The conduct of enquiry: methodology for
                                    behavioral science. (San Francisco:
                                    Chandler Pub. Co., 1964).

103   R. L. Kashyap and A. R. Rao   Dynamic stochastic models from empirical
                                      data.   (New York: Academic Press, 1976).

104   A. Kaufmann   Introduction to the theory of fuzzy
                                      subsets.   (New York: Academic Press,
                                      1975).

105   H. B. Keller and D. S. Cohen   Some positive problems suggested by
                                      nonlinear heat generation.   J. Math.
                                      Mech. 16 (1967) 1361.

106   M. G. Kendall and A. Stuart   The advanced theory of statistics.
                                      (New York: Hafner Pub. Co., 1963, 1966,
                                      1967) 3 vols.

107   C. W. Kilmister and J. E. Reeve, Rational mechanics.   (New York:
                                      American Elsevier, 1966).

108   A. Kuipers   Model and insight.   In [68] p. 125.

109   B. Lavenda, G. Nicolis and M. Herschkowitz-Kaufman, Chemical
                                      instabilities and relaxation oscilla-
                                      tions.   J. Theor. Biol. 32 (1971) 283.

110   P. F. Lazarsfeld (ed.)   Mathematical thinking in the social
                                      sciences.   (New York: Russell and
                                      Russell, 1969).

111   P. F. Lazarsfeld and N. W. Henry (eds.), Readings in mathematical
                                      social science.   (Cambridge, Mass.:
                                      M.I.T. Press, 1966).

112   W. H. Leatherdale   The role of analogy, model and metaphor
                                      in science.   (Amsterdam and New York:
                                      North Holland/American Elsevier, 1974).

113   G. Levine and C. J. Burke   Mathematical model techniques for
                                      learning theory.   (New York: Academic
                                      Press, 1972).

114   R. Levins   Evolution in changing environments.
                                      (Princeton: Princeton Univ. Press, 1968)

115   T-Y Li and J. A. Yorke   Period three implied chaos.   A.M.S.
                                      Monthly 82 (1975) 985.

116   M. J. Lighthill   Keynote address to Int. Commission of
                                      Math. Instruction.   Bull. I.M.A. 12
                                      (1976) 87.

117   D. V. Lindley   Introduction to probability and
                                      statistics.   (Cambridge: Cambridge Univ.
                                      Press, 1965) 2 vols.

118   Y. A. Liu and L. Lapidus   Observer theory for lumping analysis
of monomolecular reaction systems.
A.I.Ch.E. J. 19 (1973) 467.

119   E. N. Lorentz   The problem of deducing the climate
from the governing equations. Tellus
16 (1964) 1.

120   R. M. Loynes   The role of models. In [96] p. 547.

121   R. D. Luce and H. Raiffa   Games and decisions. (New York: John
Wiley, 1957).

122   D. Luss   Some further observations concerning
multiplicity and stability of distributed
parameter systems. Chem. Eng. Sci. 26
(1971) 1713.

123   D. Luss and P. Hutchinson   Lumping of mixtures with many parallel
first order reactions. Chem. Eng. J.
1 (1970) 129.

124   D. Luss and C. A. Pikios   Isothermal concentration oscillations
on catalytic surfaces. Chem. Eng. Sci.
32 (1977) 191.

125   E. McMullin   What do models tell us?. In [186].

126   E. J. McShane   Stochastic calculus and stochastic
models. (New York: Academic Press,
1974).

127   D. P. Maki and M. Thompson   Mathematical models and applications
(with emphasis on the social, life and
management sciences). (Englewood
Cliffs: Prentice Hall, 1973).

128   J. E. Marsden and M. McCracken, The Hopf bifurcation and its applications (Appl. Math. Sci. 19).
(Heidelberg: Springer Verlag, 1976),

129   R. M. May   Simple mathematical models with very
complicated dynamics. Nature 261
(1976) 459.

130   R. M. May and G. F. Oster   Bifurcations and dynamic complexity in
simple ecological models. Amer.
Naturalist 110 (1976) 573.

131   D. H. Mellor   Models and analogies in science. Duhem
vs. Campbell. Isis 59 (1968) 282.

132   H. D. Mellor   The matter of chance. (Cambridge:
Cambridge Univ. Press, 1971).

133   E. Nagel, P. Suppes and A. Tarski (eds.), Logic, methodology and
                                  philosophy of science. (Stanford:
                                  Stanford Univ. Press, 1962).

134   C. V. Negoita and D. A. Ralescu, Applications of fuzzy sets to systems
                                  analysis. (New York: John Wiley, 1975).

135   G. Nicolis and J. Portnow   Chemical oscillations. Chem. Rev. 73
                                        (1973) 365.

136   B. Noble   Applications of undergraduate mathematic
                                        in engineering. (New York: Macmillan,
                                        1967).

137   G. C. Nooney   Mathematical models, reality and results
                                        J. Theoret. Biol. 9 (1965) 239.

138   U. Norlén   Simulation model building. (New York:
                                        John Wiley, 1975).

139   M. F. Norman   Markov processes and learning models.
                                        (New York: Academic Press, 1972).

140   O. Ore   Graphs and their uses. (New York,
                                        Random House, 1963).

141   G. F. Oster, A. S. Perelson and A. Katchalsky, Network thermodynamics:
                                  dynamic modelling of biophysical sys-
                                  tems. Quant. Rev. Biophys. 6 (1973) 1.

142   H. G. Othmer   On the temporal characteristics of a
                                        model for the Zhabotinskii-Belousov
                                        reaction. Math. Biosci. 24 (1975) 205.

143   E. Parzen   Modern probability theory and its
                                        applications. (New York: John Wiley,
                                        1960).

144   R. Penrose   The rôle of aesthetics in pure and
                                        applied mathematical research. Bull.
                                        I.M.A. 10 (1974) 266.

145   G. Polya   How to solve it. (Princeton:
                                        Princeton Univ. Press, 1957). 2nd Edn.
                                        Doubleday Co., Inc., New York.

146   G. Polya   Mathematics of plausible reasoning.
                                        I. Induction and analogy in mathematics
                                        II. Patterns of plausible inference.
                                        (Princeton: Princeton Univ. Press, 1954)
                                        2 vols.

147   G. Polya   Mathematical discovery. (New York:
                                        John Wiley, 1962, 1965) 2 vols.

148   N. V. Prabhu

Queues and inventories. (New York: John Wiley, 1965).

149   A. Rapaport

N-person game theory: concepts and applications. (Ann Arbor: Univ. of Mich. Press, 1970).

150   A. Rapaport

Two person game theory: the essential ideas. (Ann Arbor: Univ. of Mich. Press, 1966).

151   W. H. Ray and J. R. Barney

The application of differential game theory to process control problems. Chem. Eng. J. 3 (1972) 237.

152   W. Regenass and R. Aris

Stability estimates for the stirred tank reactor. Chem. Eng. Sci. 20 (1965) 60.

153   W. T. Reid

Anatomy of the ordinary differential equation. Amer. Math. Monthly 82 (1975) 971.

154   M. Rosenblatt

Random processes. (Heidelberg: Springer Verlag, 1974).

155   A. Rosenblueth and N. Wiener

The roles of models in science. Phil. of Sci. 12 (1945) 316.

156   D. E. Rosner

The treatment of jump-conditions at phase boundaries and fluid dynamical discontinuities. Chem. Eng. Educ. 10 (1976) 190.

157   O. E. Rössler

Chaotic behavior in simple reaction systems. Z. Naturforsch. 31a (1976) 259.

158   M. F. Rubinstein

Patterns in problem solving. (Englewood Cliffs: Prentice Hall, 1975).

159   L. A. Segel

Simplification and scaling. SIAM Review 14 (1972) 547.

160   J. Serrin

Mathematical principles of classical fluid mechanics. Handbuch der Physik, Bd VIII/1. Eds. S. Flugge and C. Truesdell. (Berlin: Springer, 1959).

161   M. J. Sewell

Some mechanical examples of catastrophe theory. Bull. I.M.A. 12 (1976) 163.

162   J. A. Shohat and J. D. Tamarkin, The problem of moments. Amer. Math.
                                  Soc. (Math Surveys 1). (New York, 1943)

163   M. Sheintuch and R. A. Schmitz, Oscillations in catalytic reactions.
                                  Cat. Rev. (1977), to appear.

164   J. M. Smith   Mathematical ideas in biology.
                                  (London: Cambridge Univ. Press, 1968).

165   J. M. Smith   Models in ecology. (Cambridge:
                                  Cambridge Univ. Press, 1974).

166   T. T. Soong   Randon differential equations in science
                                  and engineering. (New York: Academic
                                  Press, 1973).

167   M. Spector   Models and theories. Brit. Jnl. for the
                                  Phil. of Sci. 16 (1965) 121.

168   D. J. Stewart (ed.)   Automation theory and learning systems.
                                  (Washington, D.C.: Thompson Book Co.,
                                  1967).

169   P. Suppes   A comparison of the meaning and uses of
                                  models in mathematics and the empirical
                                  sciences. In [68] p. 163.

170   J. Swanson   On models. Brit. Jnl. for the Phil. of
                                  Sci. 17 (1966) 297.

171   R. D. Tanner   Identification, hysteresis and
                                  discrimination in enzyme kinetic models.
                                  A.I.Ch.E. J. 18 (1972) 385.

172   A. Tarski   A general method in proofs of undecida-
                                  bility. In [173].

173   A. Tarski, A. Mostowski and R. M. Robinson (eds.), Undecidable
                                  theories. (Amsterdam: North Holland
                                  Pub. Co., 1953).

174   G. I. Taylor   Dispersion of a soluble matter in solven
                                  flowing slowly through a tube. Proc.
                                  Roy. Soc. A219 (1953) 186.

175   G. I. Taylor   Dispersion of salts injected into large
                                  pipes or the blood vessels of animals.
                                  Appl. Mech. Rev. 6 (1953) 265.

176   D. W. Theobald   Models and methods. Phil. 39 (1964)
                                  260.

177   R. Thom   Structural stability and morphogenesis.
                                  (Reading: Benjamin, 1975).

176

178   J. R. R. Tolkien   Tree and leaf. (Boston: Houghton and
                                    Mifflin, 1965).

179   C. A. Truesdell and R. Toupin The classical field theories. Handbuch
                                     der Physik, Bd. III/1. S. Flugge (ed.).
                                     (Berlin: Springer, 1960).

180   M. L. Tsetlin   Automaton theory and modelling of
                                    biological systems. (New York:
                                    Academic Press, 1973).

181   Ya. Z. Tsypkin   Foundations of the theory of learning
                                    systems. (New York: Academic Press,
                                    1974).

182   J. J. Tyson   The Belousov-Zhabotinskii reaction.
                                    Lec. Notes in Biomath. 10. (Heidelberg:
                                    Springer Verlag, 1976).

183   A. Uppal, W. H. Ray and A. Poore, On the dynamic behavior of continuous
                                    stirred tank reactors. Chem. Eng. Sci.
                                    29 (1974) 967.

184   A. Uppal, W. H. Ray and A. Poore, The classification of the dynamic
                                    behavior of continuous stirred tank
                                    reactors--influence of reactor
                                    residence time. Chem. Eng. Sci. 31
                                    (1976) 205.

185   J. B. Ubbink   Model, description and knowledge. In
                                    [68] p. 178.

186   B. Van Rootselaar and J. F. Staal, Logic, methodology and philosophy
                                    of science. III. Proceedings of the
                                    third international congress for logic,
                                    methodology and philosophy of science.
                                    (Amsterdam, 1967).

187   J. Von Neumann and O. Morgenstern, Theory of games and economic
                                    behavior. (Princeton: Princeton Univ.
                                    Press, 1947).

188   J. Villadsen and W. E. Stewart, Graphical calculation of multiple
                                    steady states and effectiveness factors
                                    for porous catalysts. A.I.Ch.E. J.
                                    15 (1969) 28.

189   J. Wei   Least squares fitting of an elephant.
                                    Chemtech. 5 (1975) 128.

190   J. Wei and J. C. W. Kuo   A lumping analysis in monomolecular
                                    reaction systems. Ind. Eng. Chem.
                                    Fundamentals 8 (1969) 114.

191   J. Williams   The compleat strategyst. (New York:
                                      McGraw Hill, 1954).

192   H. Wold   Bibliography on time series and
                                      stochastic processes. (Edinburgh:
                                      Oliver and Boyd, 1966).

193   E. Wong   Stochastic processes in information
                                      and dynamical systems. (New York:
                                      McGraw Hill, 1971).

194   J. Worster   A discovery in analysis aided by a
                                      computer. Bull. I.M.A. 9 (1973) 320.

195   L. A. Zadeh   Fuzzy sets. Information and Control 8
                                      (1965) 338.

196   L. A. Zadeh, K-S Fu, K. Tanaka and M. Shimura (eds.), Fuzzy sets and
                                      their applications to cognitive and
                                      decision processes. (New York:
                                      Academic Press, 1975).

197   E. C. Zeeman   Catastrophe theory. Sci. Amer. 234,
                                      No. 4 (1976) 65.

198   J. A. Andrews and R. R. McLone, Mathematical modelling. (London:
                                      Butterworths, 1976).

199   X. J. R. Avula (ed.)   Proceeding of the first international
                                      conference on mathematical modelling.
                                      Univ. of Missouri, Rolla (1977) 5 vols.

200   H. T. Banks   Modelling and control in the biomedical
                                      sciences. (New York: Springer Verlag,
                                      1975).

201   H. A. Becker   Dimensionless parameters: theory and
                                      methodology. (New York: John Wiley,
                                      1976).

202   D. F. Boucher and G. E. Alves Dimensionless numbers. Chem. Eng. Prog.
                                      55 (1959) 55.

203   P. W. Bridgeman   Dimensional analysis. (New Haven:
                                      Yale Univ. Press, 1922).

204   T. Bröcker   Differentiable germs and catastrophes.
                                      London Math. Soc. Lec. Note Ser. 17.
                                      (Cambridge: Cambridge Univ. Press, 1975)

205   D. R. J. Chillingworth   Differential topology with a view to
                                      applications. (London: Pitman, 1976).

206   D. S. Cohen and A. Poore   Tubular chemical reactors: the "lumping approximation" and bifurcation of oscillatroy states. SIAM J. on App. Math. 27 (1974) 416.

207   W. J. Duncan   Physical similarity and dimensional analysis. (London: Edward Arnold, 1953).

208   R. Esnault-Pelterie   L'analyse dimensionelle. (Paris: Ed. F.-Range, 1945).

209   C. M. Focken   Dimensional analyses and their applications. (London: Edward Arnold, 1953).

210   W. N. Gill, E. Ruckenstein and H. P. Hsieh, Homogeneous models for porous catalysts and tubular reactors with heterogeneous reactions. Chem. Eng. Sci. 30 (1975) 685.

211   J. Guckenheimer, G. Oster and A. Ipaktchi, The dynamics of density dependent population models. J. Math. Biol. 4 (1977) 101.

212   A. A. Gukhman   Introduction to the theory of similarity. (New York: Academic Press, 1965).

213   R. Haberman   Mathematical models. (Englewood Cliffs: Prentice-Hall, 1977).

214   J. L. Howland and C. A. Grobe   A mathematical approach to biology. (Lexington, Mass.: Heath, 1972).

215   H. E. Huntley   Dimensional analysis. (Rinehart, 1951).

216   D. C. Ipsen   Units, dimensions and dimensionless numbers. (New York: McGraw-Hill, 1960).

217   E. W. Jupp   An introduction to dimensional method.

218   S. J. Kline   Similitude and approximation theory. (New York: McGraw-Hill, 1965).

219   N. Koppel and L. N. Howard   Plane wave solutions to reaction-diffusion equations. Stud. Appl. Math. 52 (1973) 291.

220   F. W. Lanchester   Theory of dimensions and its applications for engineers. (London: Crosby, Lockwood, 1936).

221   H. Langhaar   Dimensional analysis and the theory of models. (New York: Wiley, 1951).

222   T-Y Li and J. A. Yorke   The "simplest" dynamical system.   In
                                    Dynamical Systems, Vol. 2.   (New York:
                                    Academic Press, 1976).

223   C. C. Lin and L. A. Segel   Mathematics applied to deterministic
                                    problems in the natural sciences.
                                    (New York: Macmillan, 1974).

224   K. Z. Lorenz   Fashionable fallacy of dispensing with
                                    description.   Naturwiss. 36 (1973) 1.

225   Y-C Lu   Singularity theory and an introduction
                                    to catastrophe theory.   (New York:
                                    Springer Verlag, 1976).

226   V. L. Makarov and A. M. Rubinov, Mathematical thoery of economic
                                    dynamics and equilibria.   (New York:
                                    Springer Verlag, 1977).

227   T. Maruyama   Stochastic problems in population
                                    genetics.   (New York: Springer Verlag,
                                    1977).

228   C. A. Mihram   A closer look at Thom's catastrophe
                                    theory.   Proc. 1st Int. Conf. on Math.
                                    Modelling.   X. J. R. Avula (ed.) Vol. 1
                                    (1977) 359.

229   G. Murphy   Similitude in engineering.   (New York:
                                    Ronald Press, 1950).

230   J. Palacios   Dimensional analysis.   (New York:
                                    Macmillan, 1964).

231   R. C. Pankhurst   Dimensional analysis and scale factors.

232   J. Pawlowski   Die Ähnlichkeitstheorie in der
                                    physikalisch-technischen Forchung.
                                    (Heidelberg: Springer Verlag, 1971).

233   A. W. Porter   The method of dimensions.   (London:
                                    Methuen, 1946).

234   T. Poston and I. N. Stewart   Taylor expansions and catastrophes.
                                    (London: Pitman, 1976).

235   T. Poston and I. N. Stewart   Catastrophe theory and its applications:
                                    an advanced outline survey.   (London:
                                    Pitman, 1977).

236   L. Rayleigh   Scientific papers.   (Cambridge:
                                    Cambridge Univ. Press, 1899-1919). 1, 377;
                                    4, 452; 6, 300.

237   M. Reiner   The Deborah number. Physics Today
                                       Jan. '64 (1964) 62.

238   H. Rouse   Compendium of Meterology. T. F. Malone
                                       (ed.). (Boston: Amer. Met. Soc., 1951).

239   L. I. Sedov   Similarity and dimensional methods in
                                       mechanics. (New York: Academic Press,
                                       1959).

240   L. A. Segel   Mathematics applied to continuum
                                       mechanics. (New York: Macmillan, 1977).

241   R. Shinnar and P. Naor   Residence time distribution in systems
                                       with internal reflux. Chem. Eng. Sci.
                                       22 (1967) 1369.

242   R. Shinnar, F. J. Krambeck and S. Katz, Interpretation of tracer
                                       experiments in systems with fluctuating
                                       throughput. Ind. Eng. Chem. Funda-
                                       mentals 8 (1969) 431.

243   R. Shinnar, P. Naor and S. Katz, Interpretation and evaluation of
                                       multiple tracer experiments. Chem.
                                       Eng. Sci. 27 (1972) 1627.

244   R. Shinnar, D. Glasser and S. Katz, The measurement and interpretation
                                       of contact time distributions for
                                       catalytic reactor characterization.
                                       Ind. Eng. Chem. Fundamentals 12 (1973)
                                       165.

245   R. Shinnar and Y. Zvirin   A comparison of lumped-parameter and
                                       diffusional models describing the
                                       effects of outlet boundary conditions
                                       on the mixing in flow systems. Water
                                       Res. 10 (1976) 765.

246   R. Shinnar and Y. Zvirin   Interpretation of internal tracer
                                       experiments and local sojourn time
                                       distributions. Int. J. of Multiphase
                                       Flow 2 (1976) 495.

247   V. E. J. Skoglund   Similitude: theory and applications.
                                       (New York: Int. Textbooks, 1967).

248   D. L. Solomon and C. Walter (eds.), Mathematical models in biological
                                       discovery. (New York: Springer Verlag,
                                       1977).

249   S. Smale   A mathematical model of two cells via
                                       Turing's equation. Lec. on Math. in
                                       Life Sciences. Amer. Math. Soc. 6
                                       (1971) 17.

250   R. Thom   Structural stability, catastrophe theory and applied mathematics. SIAM Rev. 19 (1977) 189.

251   J. M. T. Thompson   Catastrophe theory and its role in applied mechanics. Theor. and App. Mech. (Proc. 14th IUTAM Congress Delft. Ed. W. T. Koiter) (1976) 451.

252   J. M. T. Thompson and G. W. Hunt, The instability of evolving systems. Interdisciplinary Sci. Revs. 2 (1977) 240.

253   C. Van Heerden   Autothermic processes. Ind. Eng. Chem. 45 (1953) 1242.

254   A. T. Winfree   Spiral waves of chemical activity. Science 175 (1972) 634.

255   A. E. R. Woodcock and T. Poston, A geometrical study of the elementary catastrophes. (New York: Springer Verlag, 1974).

256   L. C. Woods   The art of physical and mathematical modelling. In A spectrum of mathematics. K. Butcher (ed.) 142. (Auckland: Auckland Univ. Press, 1971).

257   J. M. Ziman   Mathematical models and physical toys. Nature 206 (1965) 1187.

258   N. R. Amundson and O. Bilous   Chemical reactor stability and sensti sensitivity. A.I.Ch.E. J. 1 (1955) 513.

259   G. Birkhoff   Hydrodynamics: a study in logic, fact and similitude. (Princeton: Princeton Univ. Press, 1960).

260   N. Campbell   Dimensional analysis. Phil. Mag. 47 (1924) 481.

261   J. Villadsen and M. L. Michelsen, Solution of differential equation models by polynomial approximation. (Englewood Cliffs: Prentice Hall, 1978).

Subject index

Name index